# Edward Duffield

*Philadelphia Clockmaker, Citizen, Gentleman 1730 – 1803*

[ Elected to the American Philosophical Society, 8 March 1768 ]

Copyright © 2024 by the American Philosophical Society.
All rights reserved.

ISBN: 978-1-60618-009-9

Library of Congress Cataloging-in-Publication Data

Names: Frishman, Bob, author. | American Philosophical Society, issuing body.

Title: Edward Duffield : Philadelphia clockmaker, citizen, gentleman 1730–1803 / Bob Frishman

Description: Philadelphia, Pa.: American Philosophical Society Press, 2024. | Includes bibliographical references and index.

Library of Congress Control Number: 2024935826

ISBN: 9781606180099 (hardback)

ISBN: 9781606180105 (ebook)

Summary: "This book details the life and work of Edward Duffield, a clockmaker in eighteenth-century Philadelphia. The present volume is more than a comprehensive catalogue of Duffield's clocks and scientific instruments. It places his life and work in the context of the competitive environment in which he operated"-- Provided by publisher.

LC record available at https://lccn.loc.gov/2024935826

*Jacket front:* Edward Duffield standing timepiece, with spherical moon lunar indicator, Catalogue No.4. Collection of the American Philosophical Society. Photo by John Wynn.

*Jacket back:* Same clock's movement and rear of dial. Photo by John Wynn.

*Copyediting:* Lauren Byrne
*Design:* Phil Lajoie, Incollect
*Production:* Incollect, Woburn, Mass.
Printed in the Republic of Korea.
The body text of this book is set in Adobe Jensen Pro and the Chapter headlines are set in Optima.

# Edward Duffield

## *Philadelphia Clockmaker, Citizen, Gentleman 1730–1803*

Bob Frishman

AMERICAN PHILOSOPHICAL SOCIETY PRESS
PHILADELPHIA

## ABOUT THE AUTHOR

Bob Frishman was introduced to horology on Thanksgiving Day, 1980, when he was invited into the overflowing basement of a collector and dealer of antique clocks, watches, tools, and machinery. Had Bob stayed home that day, or not left the holiday dining table and gone down those stairs, this book would not have been written.

Nor would Bob's other horological efforts during the past four decades ever have happened: eight thousand mechanical clocks repaired; two thousand antique clocks and watches restored and sold; hundred-plus articles and reviews published; hundred-plus in-person and virtual lectures delivered to horological and general audiences here and abroad; annual NAWCC symposia organized at the Winterthur Museum, the Museum of Fine Arts, Boston, the Henry Ford Museum, and the Museum of the American Revolution; and exhibits created and mounted by him at venues including the Horological Society of New York and the Willard House & Clock Museum.

*Photo by Jeanne Schinto.*

Bob is a Silver Star Fellow of the National Association of Watch and Clock Collectors, and a Liveryman of the Worshipful Company of Clockmakers in London. As a dedicated supporter of other venerable cultural institutions, he is a Proprietor of the Boston Athenaeum, holder of Share Number 8 of the Library Company of Philadelphia, a member of the Grolier Club, the American Antiquarian Society, and the Ross Society of the Museum of Fine Arts, Boston.

He continues to operate Bell-Time Clocks in Andover, Massachusetts, where he lives with his wife, author Jeanne Schinto.

TABLE OF CONTENTS

*Preface*     Edward W. Kane     *viii*

*Foreword*     Jay Robert Stiefel     *x*

*Illustrations*     *xii*

INTRODUCTION     1

CHAPTER ONE     7
Edward Duffield and His Family

CHAPTER TWO     21
Edward Duffield: Merchant, Engraver, Clockmaker and Watchmaker

CHAPTER THREE     35
Edward Duffield Clock Dials and Movements

CHAPTER FOUR     49
Edward Duffield and Early Philadelphia Horology

CHAPTER FIVE     67
Edward Duffield: Citizen and Anglican

CHAPTER SIX     79
Attire of Edward Duffield and His Fellow Philadelphians

CHAPTER SEVEN     87
Edward Duffield's Philadelphia

TABLE OF CONTENTS

CHAPTER EIGHT                                                              105
Edward Duffield's Properties and Lands

CHAPTER NINE                                                               115
Edward Duffield and Benjamin Franklin

APPENDIX I                                                                 129
Edward Duffield Will

APPENDIX II                                                                132
1803 Estate Inventory

APPENDIX III                                                               140
Edward Dent's 992 Watch Parts

SELECTED BIBLIOGRAPHY                                                      143

ENDNOTES                                                                   161

ACKNOWLEDGMENTS                                                            171

ILLUSTRATED CATALOGUE
Introduction                                                               173
Clocks with Sarcophagus, Pagoda, and Arched Tops                           178
Clocks with Broken-Arch and Scroll Tops                                    192
Clocks with Flat Tops, Dwarf Clock, Bracket Clock                          226
Movements with Dials, Compasses, Sundials                                  234

INDEX                                                                      240

*Detail from An east prospect of the city of Philadelphia; taken by George Heap from the Jersey shore, under the direction of Nicholas Scull surveyor general of the Province of Pennsylvania / engrav'd by T. Jefferys, 1768, showing the area of Edward Duffield's home and shop in 1752. Courtesy of the Library of Congress.*

# PREFACE

I first became interested in clocks when I retired from a fairly intense career in business. Having more time on my hands, I found myself aimlessly wandering around various museums and I was always drawn to their galleries of antique furniture.

At the Met, MFA, Winterthur, and the top floor of the U.S. State Department, I noticed clocks inscribed by William Claggett of Newport, Rhode Island. Because my wife and I have a summer house in Newport, those clocks caught my attention but I was curious as to why I hadn't heard of him. Upon returning to Newport, I realized that almost no one in Newport had ever heard of him either. It appeared that he was better known in New York, Boston, Wilmington, and Washington than where he had lived, worked, and is buried. Thus began a long journey of researching William Claggett, learning a modest amount about clocks, and purchasing three Claggett clocks.

I then moved on to other clockmakers. This is the fourth book on clockmakers in which I have been involved as both a researcher and funder. They include books on David and Benjamin Rittenhouse, William and Thomas Claggett, James Wady, Edward Duffield, and members of the Mulliken Family. I had no involvement with a book on the Stretch family but I mention it because it is available and because it serves as a model for the rest of what has become almost a series. Interestingly, Edward Duffield is the only one of this cohort who does not appear to have been part of a clockmaker family.

My main area of interest in researching early American tall-case clocks has focused upon locating these clocks and identifying them, and I do not pretend to be knowledgeable about movements, cases, or technical matters. There already exist several well-developed lists and databases of such clocks. For example, the late Ian Quimby located a number of Edward Duffield clocks and described them in a wonderful unpublished manuscript at Winterthur. It is amazing that Quimby was able to locate several examples in those pre-internet days. A few other such lists are available online via the Yale Rhode Island Furniture Archive and the Winterthur Decorative Arts Photographic Collection. These provide much-needed points of reference and are extremely useful, but many are out of date and needed revisions as provided in this book's Catalogue.

In the process of attempting to find and identify these clocks, I noted two major themes. First, the sources of information as to possible locations were relatively consistent across almost all early clockmakers. Approximately twenty-five percent of the clocks were located using the internet which often would refer to auction catalogues, museums, magazine articles, and so forth. Another twenty-five percent emerged from discussions with collectors, dealers, and curators. Dealers often are incredibly protective regarding the identity of their clients, so it frequently took quite a bit of interrogation. Another thirty percent were identified in various reference books such as *The Clock Book* by Wallace Nutting, *Pennsylvania Clockmakers Watchmakers and Allied Crafts* by James Whisker, and *Honors's Blue Book* by William Honor, Jr.

The final twenty percent came from clocks stumbled upon by sheer luck. For example, a colleague suggested that I visit the rare-book section of the Free Library of Philadelphia in search of a David Rittenhouse clock that I hoped to find. It was not there but I discovered a magnificent Edward Duffield clock of which I was unaware.

I find the second major theme particularly fascinating. Approximately half of the total number of clocks by these makers are held by public institutions such as museums, universities, historical societies, and libraries. The other half are in private hands. That this separation of ownership is so apparently consistent across the various clockmakers seems to defy analysis. Why isn't it occasionally 90–10 or 80–20 or even 70–30? Moreover, the types of ownership appear to remain static over time.

There well may be private individuals who feel that if a museum has a maker's clock on display then that is good enough for them to consider ownership in their personal collection. It certainly worked for me regarding my purchase of three Claggett clocks. Conversely, there are curators who may feel that if a significant number of important collectors own clocks by a particular maker, then their museum should have representation. Certainly, museums occasionally deaccession and private owners donate clocks to institutions by outright gift or bequest. But this flow back and forth tends to account for a very small percentage of clocks in our universe. Moreover, they seem to offset each other. I will leave it to the reader to try to explain this unusual statistical anomaly.

I hope that you enjoy reading about Edward Duffield and his bountiful creations. I want to thank Bob Frishman for undertaking this monumental project and making it happen.

Edward W. Kane
*Concord, Massachusetts*

# FOREWORD

*I send a Book on Mr. Harrison's Watch. Present it from me to our ingenious Friend Mr. Duffield, with my love to them and their Children.*

— Benjamin Franklin letter to his wife Deborah, June 21, 1767, forwarding to Edward Duffield, *The Principles of Mr. Harrison's Timekeeper, with plates of the same;* published that year by order of the Commissioners of the Longitude.

Franklin's "ingenious Friend" Edward Duffield is the least known of three prominent clockmakers who held the most coveted horological position in eighteenth-century Philadelphia: Keeper of the State House Clock. The clock was completed and installed in 1753 by Thomas Stretch, the first keeper. In 1762 the Assembly appointed Duffield to succeed him. And in 1775, when Duffield indicated his desire to retire to the country, David Rittenhouse applied for, and got, the position. Magisterial biographies of Rittenhouse and the Stretch family have recently appeared. Now Duffield gets his due.

Franklin so esteemed Edward Duffield's friendship and ability that he appointed him a co-executor of his estate. He and his family had enjoyed hospitality at Duffield's country seat, Benfield, during some of the Revolution's bleakest days. It was there, in June 1776, that Franklin attended the first meeting of the Committee of Five appointed by Congress to compose the Declaration of Independence. Franklin's family would later shelter at Benfield during the British occupation of Philadelphia.

Duffield's horological talents were recognized by Philadelphia's scientific community. In 1769, a year after being elected its member, the American Philosophical Society hired Duffield to construct the highly accurate timepiece needed for the observation of the November 9 transit of Mercury. Duffield supplied the single–train movement with deadbeat escapement (Cat. No. 3) in under three weeks!

That timepiece and two other clocks by Duffield are preserved in the Society's collection. One, in a diminutive walnut tall case, descended in the Franklin/Bache family (Cat. No. 4). Its time-only eight-day movement has a spherical moon in its dial. This is a rare feature which the author of this volume uniquely associates with Duffield, finding it in three of his other clocks, but in none of any other eighteenth-century American clockmaker.

Several Duffield clocks are on public display in other museums, historical societies, and libraries. That at the Library Company of Philadelphia (Cat. No. 2) was described in my *Antiquarian Horology* article on the library's extensive collection of clocks and horological books. Many others are illustrated here for the first time.

Most of Duffield's surviving clocks are housed in elegant tall cases. The tallest could stand on stairwell landings, enabling simultaneous viewing of their dials from floors above and beneath. One such example, the nearly ten-foot tall "Wright Family Clock" in its rococo-carved walnut case, was photographed nearly a century ago on the landing of the center hall stairs at Merino Hill House, the Wrights' country mansion in Monmouth County, New Jersey (Cat. No. 17). The smallest of Duffield's clocks descended in the Duffield family and is one of his most remarkable. Its compact eight-day, single-train, non-striking movement with deadbeat escapement is housed in a dwarf bombé case (Cat. No. 60).

The present volume is more than a comprehensive catalogue of Duffield's clocks and scientific instruments. It places his life and work in the context of the competitive environment in which he operated. Public and private records, newspaper advertisements, even the ledger of Duffield's tailor, were mined for any scrap of relevant information. We find that, unlike most of his artisanal contemporaries, Duffield's origins were anything but humble. His substantial inheritance, possessions, and income from property established him as one of Philadelphia's most prosperous citizens. He served on civic and church bodies, and generously supported charities. At Franklin's request, in 1768, Duffield became one of two administrators of the Bray School, established to educate Black children.

With over forty years as a horologist and some 140 published articles, Bob Frishman is abundantly qualified to relate Duffield's story. This handsome volume is also a tribute to the dedication which Edward Kane, as benefactor, and the American Philosophical Society Press, as publisher, have shown to preserving and disseminating American history and material culture.

Jay Robert Stiefel
*Philadelphia, Pennsylvania*

## ILLUSTRATIONS

[Clocks and instruments are illustrated in the Catalogue on pages 172–239]

| | |
|---|---|
| Figure I.1 | Possible portrait of young Edward Duffield. |
| Figure I.2 | Purported likeness of Edward Duffield. |
| | |
| Figure 1.1 | Drawing of Duffield and Read homes in 1723. |
| Figure 1.2 | Portrait of Mary Parry attributed to Robert Feke. |
| Figure 1.3 | Portrait of Jacob Duché, stepson of Edward's aunt Esther. |
| Figure 1.4 | Miniature portrait of Edward Duffield, Jr., by James Peale. |
| Figure 1.5 | Duffield's grave at All Saints' Torresdale Episcopal Church. |
| Figure 1.6 | Edward Duffield Neill 1842 photographic portrait. |
| | |
| Figure 2.1 | First newspaper advertisement placed by Edward Duffield. |
| Figure 2.2 | Edward Duffield watch paper. |
| Figure 2.3 | Ad placed by Edward Duffield July 8, 1756. |
| Figure 2.4 | Duffield bill and receipt for Joseph Shippen. |
| Figure 2.5 | Benjamin Randolph receipt book entry. |
| Figure 2.6 | Advertisement placed by Edward Duffield selling a chamber clock. |
| Figure 2.7 | Side view of Duffield movement, Catalogue No.3. |
| Figure 2.8 | Advertisement placed by John Cox seeking lost Duffield watch. |
| Figure 2.9 | Photograph of former Lower Dublin Academy. |
| Figures 2.10–11 | 1757 Indian peace medal. |
| Figure 2.12 | Henry Inman, Pes-Ke-Le-Cha-Co Pawnee Chief. |
| Figure 2.13 | Kittanning Destroyed medal. |
| | |
| Figure 3.1 | Front plates of four Duffield movements. |
| Figure 3.2 | Front plates of four eighteenth-century movements by two English and two American makers. |
| Figure 3.3 | Disassembled movement of Duffield clock Catalogue No.39. |
| Figure 3.4 | List of names of movement parts shown in Figure 3.3. |
| Figure 3.5 | Disassembled movement by Joseph Harding, London. |
| Figure 3.6 | Comparisons of movement parts by Duffield and Harding. |
| Figure 3.7 | Strike sides of four Duffield movements. |
| Figure 3.8 | Time sides of four Duffield movements. |
| Figure 3.9 | Comparable brass dials by Edward and contemporaries. |
| Figure 3.10 | Twelve engraved dial bosses. |
| Figure 3.11 | Dials with spherical-moon indicators. |
| Figure 3.12 | Backs of dials with spherical-moon indicators. |
| | |
| Figure 4.1 | John Sprogell advertisement. |
| Figure 4.2 | Clockmaker's center lathe. |
| Figure 4.3 | 1785 movement formerly in a Newburyport meeting-house steeple. |

| | |
|---|---|
| Figure 4.4 | Diderot drawings of a weight-driven lantern clock with alarm. |
| Figure 4.5 | Microcosm ad. |
| Figure 4.6 | Pine-case astronomical clock by David Rittenhouse. |
| Figure 4.7 | Orrery by David Rittenhouse. |
| Figure 4.8 | Roasting jack at Hancock-Clarke House. |
| Figure 5.1 | Edward Duffield's signature on the original Deed of Settlement for the Philadelphia Contributionship for the Insurance of Houses from Loss by Fire. |
| Figure 5.2 | Receipt to Edward Duffield for book subscription. |
| Figure 5.3 | Reverend Richard Peters portrait by Mason Chamberlin. |
| Figure 5.4 | Copy of letter to Peters in Christ Church Minutes. |
| Figure 5.5 | Constable's warrant signed by Edward Duffield. |
| Figure 5.6 | Print of Walnut Street Goal [Jail] by William Birch. |
| Figure 5.7 | Edward Duffield advertisement seeking runaway servant girl. |
| Figure 5.8 | Edward Duffield's capture by British dragoons. |
| Figure 6.1 | Thomas Willing by Charles Willson Peale. |
| Figure 6.2 | Portrait of Chief Justice Thomas McKean and his Son. |
| Figure 6.3 | The Graisbury Ledger in the Reed and Forde Papers. |
| Figure 7.1 | William Birch print of Second Street, Philadelphia. |
| Figure 7.2 | J. Rogers hand-colored print of Pennsylvania State House. |
| Figure 7.3 | George Inn, David J. Kennedy watercolor. |
| Figure 7.4 | Pennsylvania Hospital by William Birch. |
| Figure 7.5 | Henry Dawkins, "The Paxton Expedition". |
| Figure 8.1 | 1759 survey of Edward Duffield's boyhood home. |
| Figure 8.2 | 1758 survey of Edward Duffield's Second Street home. |
| Figure 8.3 | Detail of 1767 property tax assessment for Edward Duffield. |
| Figure 8.4 | Edward Duffield advertisement selling a plantation. |
| Figure 8.5 | Photograph of Duffield homestead, Benfield. |
| Figures 8.6–7 | *McCulloch's Pocket Almanac for the Year 1801.* |
| Figure 9.1 | Portrait of Benjamin Franklin by David Martin. |
| Figure 9.2 | Dial of Franklin-designed clock made by John Whitehurst. |
| Figure 9.3 | Mrs. Richard Bache (Sarah Franklin) portrait by John Hoppner. |
| Figure 9.4 | Richard Bache by John Hoppner. |
| Figure 9.5a/b | 1776 *Fugio* dollar coin designed by Benjamin Franklin. |
| Figure 9.6 | Odometer or wayweiser originally owned by Franklin. |
| Figure 9.7 | Possible image of John Whitehurst in painting by Joseph Wright of Derby. |

# INTRODUCTION

*Collectors of antiques are likely to think of clocks as furniture rather than as timekeepers. They pay considerable attention to the case, but little or none to the mysterious mechanism within, though that is actually the clock, and in most cases the only part of the whole thing made by the man whose name appears on the dial. Cases were made by cabinetmakers, dials were usually imported, and the product of the clockmaker was the hidden work. It is these complex devices that concern the true clock collector, and the story of their development, and the growth of the clockmaking industry in America, is a fascinating one indeed.*

— Alice Winchester, The Magazine ANTIQUES, June 1952, p.507

This is the story of a highly skilled eighteenth-century Philadelphia clockmaker, Edward Duffield. His city and the years that he lived were among the most important in the history of Colonial America and the early United States. Appropriate for the publisher of this book, he was honored to be elected to the American Philosophical Society in March 1768, and in 1769, he was commissioned by the APS to produce a high-accuracy clock for timing that year's transit of Mercury. That floor-standing instrument remains at the APS to this day (Cat.No.3).[1]

But this is not the story of a typical "leather-apron" artisan. It is many interwoven stories about an eighteenth-century urban craftsman who was deeply involved in his community's economic, civic, and religious life. Edward Duffield was born wealthy and owned many properties, he served no known apprenticeship, he rarely advertised his wares or services, and he was part of the city's social elite, involved and participating in the civic activities that went with that status.

As a practitioner of a prestigious trade for a quarter-century before the Revolution, he provided costly clocks and horological services to his city's prosperous citizens and institutions. Unlike most contemporary artisans, he had no financial pressures to endure long hours at arduous tasks, nor was he critically impacted by economic disruptions. He could study, he could work, and he could engage in many activities without fear of destitution.

Edward Duffield's name adorns some of the finest floor-standing clocks produced in North America during the eighteenth century. Several are prominently exhibited at world-class institutions such as New York City's Metropolitan Museum of Art, the Baltimore Museum of Art, the Worcester (MA) Art Museum, the Winterthur Museum, Colonial Williamsburg, and the Library Company of Philadelphia. Three are in the APS collection along with two of his signed surveyor's compasses. Other Duffield clocks in elegant mahogany and walnut cases stand tall in house museums, historical societies, and private collections, including this author's. More than sixty are illustrated and described in the Catalogue section of this book.

Like other polymaths of his time, Edward, as we shall call him, possessed artisanal skills extending beyond those of his chosen craft. Although he never described himself as an engraver, Edward certainly had such skills. He produced dies for the first two commemorative medals struck in Colonial Pennsylvania. His engraved signature appears on all the clock dials, compasses, and the sundial—made for the latitude of Philadelphia—that are described and illustrated in

Detail from *Second Street north from Market St. with Christ Church. Philadelphia, W. Birch,* 1800. (Complete view Figure 7.1.) Courtesy of The Library Company of Philadelphia

the Catalogue. Edward also was well versed in building construction and design, perhaps learned from his grandfather's many ventures in real estate development. He owned many rented dwellings, he was a principal overseer of the building of Philadelphia's innovative new jail, and late in life he designed the new building for the Academy of Lower Dublin and his estate's barn that was a model for Pennsylvania stone barns into the next century. Only the school structure remains standing today.

But except for that school building and the solid substance of the clocks and instruments bearing his name, there are just shadows, footprints in the sand, from his life. Intensive research now and in the past has unearthed only a handful of primary materials. No personal account books, ledgers, or diaries have surfaced except for two factual reports that he wrote. Despite his long friendship with Benjamin Franklin, born twenty-four years before Edward, we have no letters between the two men, only passing mentions and occasional regards sent by Franklin and his correspondents.

We know nothing of Edward's upbringing, especially during the formative years after the sudden death of his mother and four brothers when he was just six years old. We know nothing of his education and training or why he chose the trade of clockmaker. There were no known horologists in Edward's family or ancestry. Neither of his grown sons followed in his footsteps; Benjamin became a respected physician, Edward Jr. a gentleman farmer.

**Figure I.1** Possible portrait of young Edward Duffield, location unknown, Courtesy of Frick Art Reference Library.

I can, however, propose a reasonable scenario for his horological career. Given his family's affluence, he was not going to be bound to a long apprenticeship under a demanding master. He may have had tutored schooling as did William Franklin, Benjamin's illegitimate son born shortly before Edward. During past centuries, a passion for horology has slowly or suddenly captured many people. These included young boys (Henry Ford), monarchs (King George III), and thousands of twentieth-century founders and members of the National Association of Watch and Clock Collectors (NAWCC). All were enchanted by beautifully-crafted, lifelike machines with ticking heartbeats, shining faces, and gesturing hands.

Young Edward may have been drawn to horology through a similar enchantment and intellectual curiosity, and perhaps began his career by tinkering with the watches and clocks that his family had the wherewithal to possess. He could certainly have been inspired by the wide-ranging interests of his older friend Ben Franklin and his brother-in-law, Ebenezer Kinnersley, who engaged in experiments with Franklin. This was in contrast to his father, Joseph, who was identified simply as a "yeoman" and does not appear to have had a lucrative trade or major landholdings. Edward perhaps chose clockmaking as a way to make his own way in the world as more than a well-heeled property-owner or general merchant.

No matter how Edward came to clockmaking, we know that he did. And we can assume that for him, as for most horologists of his era, repairing—not making clocks—would have consumed the bulk of his hours. In a city where many upper-class homes had clocks—often British imports that were considered superior to those locally made—and in which many upper-class citizens carried watches, cleanings and repairs would have provided far steadier work than an occasional order for a new bespoke clock. Even if more of his clocks are discovered, his sales of new ones could have been no more than two to five per year.

Regarding other aspects of Edward's life and times, we know much from public records about his substantial sources of income, his lands, buildings, and possessions. We know taxes he paid, the public and church bodies on which he served, and public-benefit lotteries and charities that he supported. Newspaper advertisements, many of which are transcribed in Chapter Four, demonstrate the offerings and proliferation of his local competitors. Church and civic records provide us with knowledge about the births, marriages, occupations, and deaths in his family. His tailor's ledger pages tell us what he wore. But we have only the vaguest hints about Edward himself. There are no confirmed descriptions of his height, weight, hair color, demeanor, voice, or character.

A possible portrait was purchased in 1921 by the Museum of Fine Arts, Boston (Figure I.1). This portrait was attributed by its dealer-seller to Swedish-born early-Philadelphia artist Gustavus Hesselius (1682–1755) and was inscribed as Edward's likeness.

Later experts rejected the artist attribution and questioned the sitter's identity which was written on the wood stretcher, not the canvas. A 1975 letter shared with me by MFA emeritus curator Jonathan Fairbanks, however, does give some credence to the portrait being Edward: "According to Miss Laura Luckey of the Department of Paintings, the portrait has been reattributed to 'British School.' She also tells us that it was at one time in the collection of Edward Duffield Ingraham, grandson of the sitter. The painting then apparently moved to the collection of Eli Holden of Philadelphia. It came to the Museum of Fine Arts from the collections of Eli Holden's heirs."[2]

Eli Holden (1817–1866) was a prominent Philadelphia watchmaker and jeweler. He would have appreciated a portrait of a well-known predecessor. Unfortunately, the painting was deaccessioned in 1986. Museum records claim that it was sold in October of that same year at a Doyle auction in New York City, but it is not listed in the auctioneer's catalogues from that month. Recent published advertisements seeking its whereabouts have been fruitless.

**Figure I.2** Purported likeness of Edward Duffield, Courtesy of a descendant.

Edward may also be the subject of a drawing shared with me by one of his descendants (Figure I.2). Family lore claims it is Edward, but this cannot be verified. It accompanied an incomplete Duffield clock movement that they owned.

Likewise, evidence about Edward's character is also sparse. Just two critical mentions of him are known and are discussed in following chapters. A 1757 letter to Franklin was written by an unknown local clockmaker who attacked Edward's abilities as a craftsman. A 1779 letter to Franklin from Edward's young son, Benjamin, lamented his father's limited education, illiberal attitudes, and cool inflexibility. The first of only two discovered letters written *to* Edward took the opposite tone. In 1765 the minister and colonial secretary, Richard Peters (1704–1776), expressed his great affection for Edward. A leader of Philadelphia's Anglican congregations, Peters was the most learned man in the colony. The second letter to Edward, more businesslike, was from Franklin's sister following her brother's death in 1790, when Edward was serving as an executor of the great man's will.

This book explores the nature of clockmaking and watchmaking in the eighteenth century. Well before Edward's maturity, the trade already relied on divisions of labor, specialist production of finished and unfinished components, and imports from England. Few American or European makers were making clocks from basic materials, and even fewer had the ability or tools to make watches from scratch. And nearly all *watchmakers* sold and serviced English-made watches, sometimes engraving their own names on the imported movements. Edward was no exception.

Edward had easy access to local and imported horological goods that were much easier to buy than to make. These were equal in quality to parts he would otherwise have cast, turned, drawn, scraped, filed, drilled, and milled by tedious hand labor. His well-made clock movements differ in no significant ways from examples by other Colonial and British tradesmen. The standard brass dials bearing his signature are indistinguishable from those fit to other makers' movements. Top-quality English dials were easily purchased and difficult to fabricate without special training, equipment, and materials. Dissimilarities among his movements, when he would more reasonably have made them the same from his own templates, can indicate multiple outside sources.

It must be kept in mind, too, that until Parliament enacted the infamous Stamp Act in 1765, there was no "Buy-American" sentiment among the colonists, all British subjects, who were living in Pennsylvania and other colonies along the Atlantic coast. In fact, the wealthiest of them had a marked preference for English-made goods. It is quite likely that, before the rise of anti-importation sentiments, Edward would have promoted his clocks to his wealthy customers as having substantial amounts of English content and looking quite like those being imported as complete clocks or as complete movements and dials ready to be cased locally. He had absolutely no incentive to produce and market clocks that did not look English or had solely local content.

At the same time, while producing an entire movement from raw materials during Edward's years was decidedly old-fashioned, he did not simply buy and sell finished products, as Winterthur curator Charles F. Hummel intimates when describing a clock in its collection (No.18 in the Catalogue): "eight-day movement *made or assembled* by Edward Duffield."[3] Instead, he was likely to have completed clocks from rough prefabricated components and standard designs, and those tasks were certainly not quick and easy. They required substantial expertise, experience, and arrays of specialized tools of the trade. George H. Eckhardt, in his classic 1955 book, *Pennsylvania Clocks and Clockmakers: An Epic of Early American Science, Industry, and Craftsmanship*, makes this clear:

> It is true that clockmakers in Pennsylvania often bought wheels, plates, and other parts which they finished and built into clocks. But this was never a matter of mass production, since every clock was an individual effort.... The fact that Pennsylvania makers often bought parts, usually imported from England, has been given too much emphasis by present-day writers. What these men bought was far indeed from a finished clock. And even though a clockmaker could buy parts his resourcefulness was often taxed.[4]

Four of Edward's clocks, discussed in Chapter Two, demonstrate his inventiveness. These four clocks have dials with spherical-moon lunar indicators. This rare feature is found on a small number of English clocks but Edward's system for rotating the metal globes is different and simpler, and suggests that he may have adapted imported English dials to his design in order to offer a unique and costly upgrade. I know of no other early American clocks with it.

It must be emphasized that Edward did not make his clocks' wooden cases. Had he penned an autobiography, it certainly would not have been a furniture book. In Edward's time, "clock" most often meant only the metal machine and dial. The case that housed them was named separately and ordered separately by the purchaser. However, Edward was likely to have collaborated in his customers' case choices and construction. He certainly needed to be sure that his parts would fit properly into the cases being crafted nearby. He may well have discussed with the clients and cabinetmakers the choices of woods, the levels of quality and ornamentation, the finishes, and the delivery and setup of the finished product. Therefore, the Catalogue section fully illustrates cases containing Edward's movements but does not address case construction and design. That was not Edward's work nor is it my expertise as a professional horologist.

The Bibliography lists excellent books describing Philadelphia clocks as furniture. For readers seeking information on cases' secondary woods, glue blocks, joints, carvings, hardware, feet, fretwork, columns, etc., these books plus many of the Catalogue's entries provide references to museum and published descriptions presented more authoritatively than possible here.

Issues of what Edward made, and did not make, should in no way diminish our appreciation for his clocks. We simply should acknowledge how he and his eighteenth-century colleagues needed to operate efficiently and profitably. Edward's name on a clock confirmed to its initial purchaser, just as it does today, that the clock was a highly valuable object that he was proud to finish, sign, and sell.

That he mostly repaired timekeepers, rather than made them, should also in no way lessen our respect for Edward's expertise. As I can verify, there is nothing easy or routine about repairing intricate mechanical machines ranging from pocket watches to room-size movements. In 1980 I was warned that I would need ten years of hard practice and costly mistakes to become a proficient repairer and that I might reach a basic level of competence when I could fix my own inevitable mistakes. Even now, the simplest overhaul can be plagued by malfunctions, errors, accidents, defective parts, lengthy troubleshooting, and dreaded "comebacks" days or months later. Edward had the same or worse repair challenges. He endured poor lighting and minimal magnification, longer supply chains, and greater variations in available metals, supplies, and parts. He had no central heating or air conditioning. Unlike repairs of most modern mechanical devices like automobiles and lawnmowers, clock and watch service requires highly skilled concentration, not just swapping bad parts for new. An old ticking clock has no output to computer diagnostics.

This book is far more than a history of eighteenth-century Philadelphia horology from one craftsman's perspective. It is the history of a prominent clockmaker, citizen, landowner, and intimate of many of the luminaries who helped shape his city and his new nation. Philadelphia's key events are discussed not only for context but because almost certainly Edward witnessed or directly participated in them. It would have been impossible for him not to, given his location, friendships, family connections, and social standing.

The following chapters focus on many aspects of Edward's life and city as well as on eighteenth-century horology. Chapter One provides the reader with an account of Edward's family background. Chapters Two through Four consider different aspects of his horological career. Chapters Five and Six look at Edward's civic and social place in his city, while Chapters Seven and Eight detail his property holdings and provide an account of his tailor's records. Finally, Chapter Nine reveals his close connections with Benjamin Franklin. Their long and close friendship further confirms the importance of Edward Duffield's story.

CHAPTER ONE

# Edward Duffield and His Family

*Benjamin Duffield was twenty-one years of age when he left England.... In 1685, he purchased a portion of the tract of land which Allan Foster, his brother-in-law, obtained in England from William Penn... Benjamin Duffield and his wife were prospered in every respect.*

— Edward Duffield Neill, 1875

Much of Edward's family history needs to be pieced together from the scant sources discussed in the Introduction. The cast of characters during a large part of Edward's life—none except Edward were associated in any way with clockmaking—includes his wife, his grandfather and father, his two sisters and their husbands, and his two sons and two daughters. As with most eighteenth-century families, many of Edward's relations, siblings, and offspring died young. His parents had nine children but only three survived them—Edward and his sisters, Elizabeth and Sarah. Edward and his wife, Catherine, had seven children: just four reached adulthood. Edward's mother died suddenly when he was six years old.

The Duffield surname can be traced back to 1315 in England and has descended through Anglo-Saxon culture in Derbyshire and North Yorkshire. Duffield, which gave Edward's family its name, is a village on the River Derwent at the mouth of the River Ecclesbourne, about 130 miles northwest of London. (Coincidently, it sits just five miles north of the town of Derby, where Franklin several times visited his British clockmaker friend John Whitehurst (1713–1788)). Other English villages bearing this name, variously spelled, appeared even earlier in the 1086 Domesday Book. One was listed as Dufeld and another as Dunelle or Duvelle, meaning open land frequented by doves.

Edward's known ancestry begins with Robert Duffield, his great-grandfather. The patriarch was born in 1610 in Yorkshire. In 1661 he fathered Benjamin who, together with his wife, Elizabeth, whom he married in England, left Hull in 1678 on *The Shield* and arrived in Burlington, New Jersey. Together the couple had thirteen children, some born in England and others in America.

Four years after his arrival in the New World, Benjamin reportedly saw William Penn step off *The Welcome* in 1682. The Duffields were not Quakers, but they were accepted into the new colony according to Penn's insistence on religious toleration.

*Dial detail of Duffield-signed bracket clock (Catalogue no.61), Collection of Winterthur Museum, Garden and Library.*

With Benjamin established in Pennsylvania, his father, Robert, at age seventy followed from England, settling on land owned by Allen Foster, who married Benjamin's sister, Mary. Benjamin himself purchased land from Foster and in September 1685, bought 111 more acres of land in Dublin Township.

Edward's father, Joseph, the eighth child of Benjamin and Elizabeth, was born in 1692, the same year that Robert died and was buried in the Pennepack Baptist Church cemetery.[5] Edward's mother, Sarah Koster, was born in 1694.

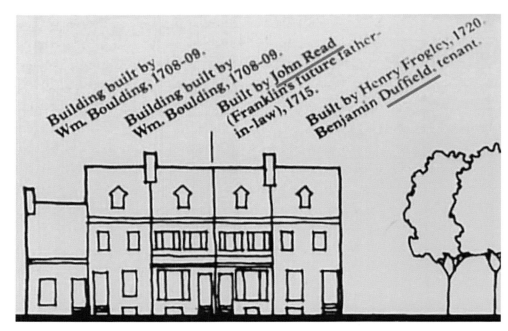

**Figure 1.1** Drawing of Duffield and Read homes in 1723. Chapter II, Illustration No.5, of National Park Service Historic Structures Report on 316 Market Street, March 1961 (author's red lines).

By 1701 Benjamin was already a key member of Philadelphia's community. That year, Lieutenant-Governor Andrew Hamilton appointed him to a commission considering a new road running north out of the city. In 1704, he acquired two houses and a tanning yard at Second and Chestnut streets. He continued his land purchases and construction of houses, shops, stables, and leather-finishing buildings, becoming one of the city's major non-Quaker property-owners.

In 1709, Benjamin's poor-tax assessment placed him in the top fifth of all local taxpayers. Investing outside the city, he purchased 600 acres in 1709 in the Manor of Moreland, which had been established by London physician Nicholas More, who arrived in 1682 with a grant from Penn for 9,815 acres.[6] Edward's future inherited estate was within this landholding. Edward Duffield Neill, a direct descendant of Edward and one of our best sources about his life, reported that Benjamin on his rural property "established a gristmill, saw-mill, tannery, and blacksmith shop."[7]

Benjamin was elected in 1711 to the first of his terms in the Pennsylvania Assembly, serving again in 1712, 1715, and 1721. He represented county voters, not the city's Quaker-controlled districts. He was not an active legislator or debater. "He played a remarkably minor role for a man of his financial significance… Duffield's increasing wealth apparently did not translate into local public service…"[8] In 1716, he moved back into the city, living on Chestnut Street, and was admitted the next year as a city freeman, formally allowing him to pursue his trades.

There are no marriage records for Edward's parents, Joseph and Sarah, but his oldest brother, Benjamin, was born on April 30, 1716, with no baptism recorded. His sister Elizabeth arrived on December 31 of the same year and was baptized on March 15, 1717. In 1718 Edward's sister Mary was born February 23 and baptized April 9. His sister Sarah arrived February 26, 1722, and was baptized April 15. In 1724, Edward's sister Hannah was born May 19 and baptized August 23. His brother Joseph was born March 12, 1726, and his brother James in 1728. All except Benjamin were baptized at the Abington Presbyterian Church. Edward was born on April 30, 1730, on the estate that he would later inherit. No baptism date is known. (His birth year has been stated as 1720 but this contradicts his gravestone, a family Bible[9], and other events of his life.) Finally, in 1733, Edward's youngest brother and final sibling, Uz, was born May 4, and his baptism may have been recorded on August 4. Indistinct records of the Presbyterian Church in Abington show more clearly his siblings' names inscribed earlier.

During this same period, Edward's grandfather, Benjamin, purchased a house at 110 High

Street (later 320 Market Street) that was built three years earlier in 1720 by Henry Frogley whose widow then sold it. Benjamin Franklin's in-laws, the Reads, lived next door at 108. That Duffield home was associated with the family until 1752 and still stands as the re-created Benjamin Franklin Printing House run by the National Park Service, although Franklin's shop was actually at another nearby location. A drawing shows these houses as they stood in 1723 (Figure 1.1).

Edward's older sister Elizabeth married Samual Swift on June 25, 1733, at Christ Church. The next year Edward's father, Joseph, owned 200 acres in the Manor of Moreland, perhaps already given to him by his father, Benjamin, who was living nearby. Before his death, Benjamin deeded to Joseph the dwelling at 110 High Street, adjacent to the Reads, where Edward most likely lived for some of his boyhood years.

The worst year of Edward's childhood certainly was 1736. Within a span of twelve days between August 23 and September 4, five family members died including all four of his brothers—Benjamin (age twenty), Joseph (age ten), James (age eight), and Uz (age three), and his mother (age forty-two). A smallpox outbreak may have been the cause. Only his father and older sisters survived to care for six-year-old Edward.

In 1739, the second of Edward's sisters, Sarah, married Ebenezer Kinnersley (1711–1778), born in Gloucester, England, and identified as a shopkeeper although he was also a trained minister. The following year Kinnersley initiated a theological dispute with his Baptist church (which was not the place of worship of his new wife or young Edward). As assistant pastor, he strongly objected to the senior pastor, Jenkin Jones, welcoming radical Presbyterian hellfire preacher John Rowland to the pulpit. Kinnersley denounced Rowland in a sermon, causing many congregation members to walk out during it. He later prevailed upon Franklin, a friend of evangelical George Whitefield, whom Kinnersley also attacked, to print that sermon. Franklin added a preface indicating that he was not taking sides. Kinnersley's unhappy schism with his church greatly affected the direction of his life and livelihood shared with Edward's sister.

Kinnersley was much admired later as an educator and he closely collaborated with Franklin's electricity experiments, perhaps with young Edward looking on. Harvard University Professor I. Bernard Cohen wrote: "But of all the lecturers on scientific subjects in Philadelphia in the middle of the eighteenth century, the most famous and most popular was Ebenezer Kinnersley … a co-experimenter of Franklin's … [he] himself made several important discoveries, such as the conducting qualities of various sorts of charcoal; he also invented an electric air thermometer."[10]

Franklin helped draft lectures that Kinnersley delivered to paying audiences in many colonial cities, and Kinnersley wrote to Franklin that the invention and use of lightning rods would "make the name of Franklin like that of Newton, *immortal*."[11] On a trip in 1749 to Virginia and Maryland, Kinnersley demonstrated an early form of electric shock therapy to cure human ailments, including "tooth ache, pains in the head, deafness, pains in the limbs… swelling of the spleen, sprains, relaxation of the nerves."[12] It is possible to imagine that his interest in experimentation inspired his young brother-in-law's interest in clockmaking.

In 1741, Edward's grandfather died May 1 at age eighty and was buried at Christ Church. The gravestone was inscribed: "The age of man is but a span, His days on earth are few; In death he must embrace ye dust, and bid this world adieu." His executors were his son Joseph, Edward Bradley (husband of his daughter Esther), and Thomas Whitton. His will does not mention his wife, Elizabeth, who likely predeceased him, but it names four children and nine grandchildren, including Edward.

At his death Benjamin held at least fifty-two mortgages totaling more than £4,600 and many deeds on which he was variously described as a "tanner, yeoman, gentleman."[13] His heirs received 1,183 acres of land plus eleven additional properties, nearly £400 in cash and wearing apparel, and

£65 in silver plate. Benjamin left his son Abraham just five shillings due to some probable dispute.

His favored grandson, Edward, received a third of the estate residue not specified in the will, extensive properties in the city and in the Manor of Moreland, and a fifth of his silver. A porringer in the collection of the Metropolitan Museum of Art[14] was claimed by its donor to have been inherited by Edward from his grandfather. The corner plot on Mulberry (later Arch) Street, Edward's future home, went to his father until the boy reached maturity.

The first mention of Edward in connection with the family of his wife, Catherine, occurs when he appears two years before their marriage as a witness to a codicil of the will of George Smith, the second husband of Catherine's mother, Mary Parry, née Humphrey. Mary had previously married David Parry on July 24, 1724, at Christ Church, and in 1727, their daughter Catherine was born October 14 and was baptized March 8 the following year by Reverend Archibald Cummings.

**Figure 1.2** Portrait of Mary Parry attributed to Robert Feke. Courtesy of private owner and descendant. Photo by Mitro Hood.

She was one of six children. All that remains from Catherine's childhood is an intricate linen sampler that she stitched at age twelve, which is in a private collection. One of the aphorisms the girl chose to include was: "This work in hand my friend may have when I am dead and laid in grave." David Parry, her father, died in 1741, the same year as Edward's grandfather, and was buried at Christ Church. The cause of neither death is known. One possibility is yellow fever, then called the Palatine distemper, and sometimes blamed on German immigrants, which had reappeared in the city for the first time since 1717 and killed 250 people. This seems unlikely, however, as Benjamin and David each died before warm days brought the mosquitoes responsible for the disease.

George Smith, Catherine's stepfather, died in 1749. His two-page estate inventory, listing contents of the house where Catherine still resided, included two unnamed clocks and cases, a jack and spit in the kitchen, and also in the kitchen four "Negro" workers, three of them elderly. Among the contents of the estate may well have been a portrait of Mary Parry Smith, which is attributed to artist Robert Feke (c.1705–c.1752), who visited Philadelphia in 1746 (Figure 1.2). Portraits by Feke were commissioned by wealthy residents including Benjamin Franklin (the first one made of him), Thomas Hopkinson, Tench Francis Sr., and Mary McCall as well as Mary Parry Smith and others.

Edward's father died in 1747. Sometime after his first wife's death, he married Hannah Leach, but when he did so and she became Edward's stepmother is unknown. His will, dated February 3, left to her all his silver and household goods as well as his "tenement and lot" on the south side of High Street "during the term of her natural life, in lieu of dower."[15] She also received an enslaved woman, Sarah. A second, named Phillis, went to Edward's sister Sarah, Kinnersley's wife. Elizabeth

Swift, Joseph's other daughter, received the yearly ground rent for property he had purchased from John Palmer. Edward's two brothers-in-law, Samuel Swift and Kinnersley, were the executors. Edward also was listed as an executor, but he did not serve, perhaps because he was underage at the time.

The executors advertised in July of the following year that persons indebted to the estate should make speedy payment or face legal action.[16] In the same notice, they offered for sale a twenty-foot by one-hundred-twenty-foot lot on the east side of Market Street between Callowhill and Vine streets adjoining a lot owned by Richard Peters, perhaps already Edward's friend. When Hannah died that year, Edward inherited his third of Joseph's remaining estate.

On June 5, 1747, Edward's aunt Esther became the second wife of Colonel Jacob Duché Sr. (1708–1788) who later sat on the Common Council, served as mayor, and was a Christ Church vestryman. At the time of Edward's marriage four years later, she paid him just twenty shillings for land on Passychurch Road in Passyhunil Township. She may have needed financial help or he may have been expressing thanks for her possible care for him after his mother died.

This link to the Duché family was one of Edward's many connections with city and church leaders, including Jacob Duché's scholarly son, Jacob Duché Jr. (1737–1798). The younger Jacob, in 1759, married the sister of Edward's close associate, Francis Hopkinson (1737–1791). Duché Jr. was a Christ Church rector and the first chaplain to the Continental Congress (Figure 1.3). However, he became a Loyalist during the British occupation of Philadelphia and was replaced in 1777 by William White and Edward's Presbyterian cousin, George Duffield. For switching allegiances, Jacob Jr. was convicted of high treason and had his estate confiscated. He fled to England but returned to Philadelphia in 1792 and may have renewed his friendship with Edward.

Edward reached age twenty-one in April 1751, and married Catherine Parry at Christ Church on June 10 of the same year, although couples at that time usually could not afford to marry so young. The average age in England for marriage was twenty-six for women and twenty-eight for men, although this was somewhat lower in America. Edward's inherited wealth enabled him to take a wife and to open a shop upon reaching maturity.

On May 11, 1752, Edward and Catherine's first child, Mary, was born "before one in the morning" and baptized July 19. The Humphrey family Bible inscribed specific times for births in Edward's and Catherine's family, and transcriptions of those pages were viewed in Washington, D.C., at the Daughters of the American Revolution Library. Those times indicate that clocks or watches were present and consulted during momentous events, and the times will be noted for other family members as well. Mary died the next year on June 7, and was buried with other family members at Christ Church. Their first son, Benjamin, was born "at half an Houre after Eleven at night" on November 3, 1753, at the Benfield estate.

That same year the Duffields sold the family home on High Street. Wigmaker George Cunningham paid £350 to Joseph Duffield's heirs, with Edward taking a £100 mortgage for his share. Cunningham at that time shaved Benjamin Franklin's head and face every two weeks.

**Figure 1.3** Portrait of Jacob Duché, stepson of Edward's aunt Esther. Unknown artist. Courtesy of Wikimedia Commons.

Edward and Catherine's daughter Sarah was born the first day of 1756 and was baptized February 20 by Reverend Robert Jenney. She later married Stacy Hepburn.

On May 17, 1757, the first graduation ceremony of the College of Philadelphia included Francis Hopkinson. Edward likely attended to witness the award of an honorary degree to his brother-in-law, Kinnersley. The following year, Kinnersley published *Further Experiments in Electricity* that described some of his research with Franklin. In 1762, Kinnersley was named the college's director for out-of-town students. He and Edward's sister oversaw the school's dormitory when it was completed the following year.

Edward's son, Joseph, was born on August 23, 1758, but that infant soon died on October 19. Three years later, Edward and Catherine's daughter Elizabeth arrived on September 10 and was christened at Christ Church on September 23. January 11, 1764, was the birthdate of their son Edward Jr. There is a 1790 miniature portrait on ivory of him at age twenty-six by James Peale of the famous artistic family (Figure 1.4).[17] The son was baptized April 13 at Christ Church by The Reverend Richard Peters.

On October 18, 1765, Edward and Catherine's final child was born, a daughter named after her mother. She died young in 1774 on August 23.

In 1756, Edward's brother-in-law, Dr. Samuel Swift, advertised the sale by public vendue of a public house and twenty-three acres in Lower Dublin township, nine miles from Philadelphia. The description offers insights into what a property close to Benfield looked like:

> The House is new built with Stone, two Story high, with good Cedars, and Plenty of Room; there is a Well of excellent Water close by the Door; it is a good Stage from Philadelphia for Travelers to stop at, and a very pleasant Ride in the Spring and Summer Seasons for Pleasure…. There has been a Store on the Premises many Years and it would be a good Place for the Business, if there was one now, with a good Assortment of wet and dry Goods; or it might suit a Gentleman for a Country Seat, to take his Family to live at in the Summer Season.[18]

**Figure 1.4** Miniature portrait of Edward Duffield, Jr., by James Peale, 1790. Location unknown. Image courtesy of Elle Shushan.

Mary Parry Smith, Edward's mother-in-law, died in March 1766. For half a century, she had been a best friend of Franklin's wife, Deborah, and their daughter, Sarah, who wrote to her father in London with the news:

> Our dear Friend Mrs. Smyth [sic] after an illness of 5 months and 6 days Expired Yesterday morning. In the whole time she had not been out of bed a quarter of an hour at a time, so thankfull she was for anything her friends did for her and patient to a Miracle. Poor Mrs. Dufield and poor Mama are in great distress, it must be hard to lose a Friend of 50 Years standing, but when we saw her in such extreame pain it would have been selfish to wish her stay, when so much happyness await'd her.[19]

Mary's will named Edward as executor and beneficiary, "in consideration of the Love & Affection which I bear unto him" and it was witnessed by Francis Hopkinson along with David and Mary Chambers. Her estate inventory included a clock and case in the parlor and another in the back room, both of which could have been provided by Edward. He himself placed a newspaper announcement: "All persons indebted to the Estate of Mary Smith, late of Philadelphia, deceased, are desired to pay to the Subscriber, without further Notice; also those who have any Demands against said Estate, are desired to bring in their Accounts properly attested, that they may be settled and paid, by Edward Duffield, Executor."[20]

A year previously, in May of 1765 Dr. John Morgan (1735–1789), recently back from training in Edinburgh, delivered a discourse about medical education at the College of Philadelphia's commencement. He proposed a new system and established America's first medical school at the College. Edward's son Benjamin, later a physician, was eleven years old, and may have been inspired either by hearing this address or reading it in its published form. In either case, three years later, Benjamin paid tuition to the College of Philadelphia and began studying medicine. The receipt was signed by Kinnersley, his uncle. In October of the next year, Benjamin began five years of medical studies at the Pennsylvania Hospital. In 1771, he received a master-of-arts degree. At the ceremony, he recited his lengthy poem, *Oration on Science*, closing with exclamations on the blessings of scientific endeavors:

All that ennobles man's exalted race,
All that Religion, Virtue, Truth, embrace!
'Tis hers with loftier feelings to inspire,
And fit a mortal for a heavenly choir![21]

He continued training with influential doctor John Redman. Although no diplomas were issued in the turbulent year of 1774, Benjamin received a certificate affirming that he had successfully completed his five-year apprenticeship and was qualified in the healing arts. Professors John Morgan, William Shippen, Adam Kuhn, and Benjamin Rush signed the document.[22]

Wealthy iron manufacturer Thomas Potts, a friend of Edward's and into whose family the young doctor later married, wrote in August to Franklin: "My friend Mr. Duffield's son Ben is going to Edenbourough [sic] to improve himself in his profession of Phisick by whom I write this and he promises to send by some friend from there. If he should ever have the pleasure of seeing you in London you will easily discover Merit in him."[23]

When Benjamin returned in 1775 from his medical training in Scotland, he became an assistant surgeon for the new Continental navy under Dr. Rush, by now one of Philadelphia's leading physicians. The following year, the two men took charge of the city hospital and pest house. Benjamin reported to the Committee of Safety, on which David Rittenhouse served, that he examined the Pest House on Province Island and found it in good order and sufficient for the accommodation of forty patients.

Philadelphia cabinetmaker Jonathan Gostelowe married a Mary Duffield on June 15 of 1768. She has been identified as Edward's niece, but he had no brother with a daughter of that name. Most likely she was a cousin, born in 1748 to Edward's uncle John, who had been a son of his father's brother Thomas.

Sadly for Gostelowe, this Mary died just two years after her wedding. She left to her widower a sizable estate of houses and land inherited from her and Edward's ancestor Benjamin. His will provided substantial properties and rents and an annual payment to his son Thomas, and "To his grandson John, he left three messuages on the south side of Chestnut Street in Philadelphia, and one fifth of his plate." With his Duffield family connections and a shop very near to Edward's, Gostelowe might have crafted cases for Edward's clocks although no cases for any maker have been attributed to him.

In August of 1770, Edward's brother-in-law Samuel Swift announced a runaway indentured joiner, George Owen, who stole many items including, "2 silver china faced watches."[24] A forty-shilling reward was promised for the return of watches that perhaps came from Edward.

In 1772, Edward was executor, with John Swift, of the will of widow Margaret Susannah Bishop of Moorland. The deceased was a friend of Edward's sister Elizabeth.

That year, Edward's cousin, Reverend George Duffield, became minister of the Pine Street Presbyterian Church in Philadelphia. The appointment was controversial due to disagreements with his doctrinal views and objections from the Second Presbytery of Philadelphia. George was in a branch of the family that did not switch to the higher-status Anglican Church.

In March 1773, Kinnersley wrote to Edward's sister from Barbados where he had gone to improve his health. He complained of lameness and of the illness of his enslaved servant, Caesar, who accompanied him. Taxes and records from the College of Philadelphia indicate that Caesar was in bondage from at least 1757 to 1774. While Kinnersley was Steward of the college dormitory, he was paid for Caesar's services including bell ringing and making fires. The two men returned home from Barbados in June when Kinnersley and Sarah's daughter Esther died. In August, he experimented on electric eels with David Rittenhouse and Owen Biddle, perhaps with Edward looking on.

Five years later in March 1778, Kinnersley died at age sixty-seven. As an executor, Edward received £500 from Philadelphia College for Kinnersley's electrical equipment appraised by David Rittenhouse. No itemized list of this equipment, nor the instruments themselves, are known today. Kinnersley had an eight-day clock perhaps supplied by Edward, valued at £30 and second only in worth to some bonds and his library of 154 volumes.

In April of the same year, Edward's son Benjamin married Rebecca Potts of Pottsgrove Manor. The Potts family descended from Thomas Potts who arrived in Philadelphia on the same ship as Edward's grandfather, which is perhaps when the families' friendship was formed. They founded colonial America's largest iron-maker. An Edward-signed tall clock (Catalogue No.1) stands in their historic house museum. On the day before the wedding, the couple signed a lengthy indenture,[25] mandating that £700 the bride received from her father's estate was for her sole use and that her new husband "shall not nor will intermeddle with, nor have any right to the same." Edward cosigned the document along with Samuel Potts and Andrew Robeson.

Soon after, on May 30, the groom joined members of the Potts clan at their Warwick Furnace in Chester County to take the Oath of Allegiance to Pennsylvania. The Test Act mandated this oath for all white male inhabitants over age sixteen,[26] so we can assume that Edward swore to it as well.

A year later, although it is unclear why, Benjamin was back in Europe in dire straits and at odds with his father, as is clear from a revealing letter he sent from Bourdeaux to Franklin in Paris:[27]

Bourdeaux May 16. (1779)

Dear Sir.

You will probably be much surpized at receiving a Letter from this Place, and from one who has so much disgraced the Introduction you were so kind as to favour him with. I cannot without deep Confusion, attempt to address you again—I cannot attempt toe excuse my past follies and Indiscretions otherwise than by pleading a natural Volatility and a great flow of Animal Spirits. These joined to Youth and Health hurried me in Imprudencies that I have since severely repented of. I have bought Experience dearly. Distress and Necessity have taught me a Lesson that I never, never shall forget: and I am now by my regular Conduct, and prudent behaviour, endeavouring to regain that Reputation that I once enjoyed. The Advice and Assistance of my amable Friend H: Conyngham have been of the utmost Service to me, and as the first stop toward reinstating myself in the good Opinion of my father, he advised me to return, & most generously furnished me with everything necessary to accomplish that Intention. I was within a few Days of sailing, when I received a small Bill from my father—with a Letter strongly marked with that cool Inflexibility for which you my dear Sir well know he is so remarkable. Indeed he has Reason to be much offended—but his Education being much confined, and his Ideas of Men and Things not the most liberal, he magnifies my offences. However; if he is implacable there is a Door open for my Reception. The Army of my Countrymen shall be my Resource; and as I am happy in an excellent Constitution and possess some degree of Firmness of Mind, I hope to bear his Slight and Displeasure with Patience and Resolution. I am already resigned to it, I am so fully convinced of having deserved it.

Mr. Conyngham (whose Name I shall never mention but with the most sincere Gratitude) has promised to to procure me a Passage in a Letter of Marque to America. I shall officiate in the way of my Profession if called upon—tho if in that Quality I could get on Board a Cargo Privateer it would be much more acceptable to me: and perhaps profitable as my finances are but slender. My Bill was for £85; Part of this I employed in liquidating some Debts that my Pride and folly led me to contract—the Remainder being 28 Guineas I have with me—and as I am in want of cloaths If I can embark either in one method or the other with a very few Guineas in my Purse I shall be contented.

If you do not think me quite unworthy your Notice—if you have any Compassion for my unhappy situation, may I request the favour of your advice. I look upon your Character as sacred, & as I am shut out from Parents and all that I hold dear, and that once held me high in their affection and Esteem, will you condescend to make out a Line for my Conduct which I promise most religiously to observe. I have seen my Errors in the strongest Light, and will endeavour to make my future Conduct as bright as my faults were deep before.

I am my dear Sr. your most obliged humble servt. B. Duffield

No reply or counsel from Franklin is recorded. This situation may in part explain why Benjamin's marriage indenture so carefully protected his wife's money. Safely back in America and having served as a surgeon during the Revolution, Benjamin in 1783 was a physician living and working on Front Street, between South and Almond streets, so he seems to have overcome whatever problems he faced earlier.

Less is known of Edward's second surviving son, Edward Jr. A number of Edward Duffields appear in local records. One was on the roll of Captain David Snyder's Light Dragoons for the County of Philadelphia, but this conceivably was a cousin. A 1786 militia roster listed another, but Edward Sr. would have been beyond the age limit of fifty-three and no longer a city resident, nor was the younger Edward. A 1786 Montgomery County militia listed an Edward Duffield as number seventy-three. This possibly was Edward's son serving in Captain Andrew Buskirk's Moreland Company, part of The Second Battalion of the Montgomery County Militia. Minutes from 1798 show the son as a 2d Lieut. in Captain John Lardner's Philadelphia County Third Troop of Light Dragoons.

Edward Jr. in 1787 was affiliated with All Saints' Church in Torresdale and was a deputy to the Diocesan Convention. His father worshipped there as well, no longer renting a prominent pew in Christ Church. In May 1789, Edward Jr. was Secretary of the Philadelphia County Society for Promotion of Agriculture and Domestic Manufacturers. He notified its members that George Washington was elected as a corresponding member.

As an adult, Edward Jr. was public-spirited like his father. Edward Duffield Neill wrote of his uncle: "He was a gentleman of refinement and of high integrity. He devoted his time to reading and to agriculture. His wisdom and judgment as a referee were sought for in every direction."[28] In 1794, he joined local citizens in establishing the new Byberry Library at the home of Ezra Townsend in Bensalem. Each subscriber agreed to pay four dollars for book purchases and an additional dollar annually to lease the space. He was one of three managers for an authorized lottery to raise $5,004 to complete the Academy of Lower Dublin, for which, as noted in Chapter Five, his father designed the school's building and was its generous supporter. Edward Jr. was a county election agent in 1802 and a vestryman at All Saints' Church.

Three years after older-son Benjamin's nuptials, there was another family wedding. Edward's daughter Sarah was married in September 1781 to Stacy Hepburn at Christ Church with Revered William White officiating. Franklin's daughter Sally wrote about this happy event to Temple, Franklin's only grandson, who was working with his grandfather in Paris at the time: "Miss Duffield is soon to the married to a Mr. Hepburn a Merchant of this place, and one that courted her more than ten years ago, she did not then fancy him, her taste is since altered and she seems now much attached to him."[29]

This wedding was followed ten years later by the marriage of Edward's daughter Elizabeth in April 1791 to Francis Ingraham at Christ Church. The Reverend William White performed the ceremony. Their son Edward Duffield Ingraham is mentioned in a postscript below.

Edward in 1793 paid £50 to his widowed sisters for five acres of land, perhaps to aid them financially and ease their later years. The indenture was witnessed by Jonathan and Edward Swift and signed before Justice Jonathan Schofield. In this document, Edward was identified as the "late Clock and Watch maker," further indicating that he no longer was in the trade.[30]

Philadelphia diarist Jacob Hiltzheimer (1729–1798)—his original diaries are in the APS Library collection—reported that on July 20 of that year he dined outdoors under trees with a group including Edward's son Benjamin. Living then at 303 South Front Street, the doctor was among the local physicians who remained in the city during the yellow fever epidemic, along with his cousin Dr. Samuel Duffield at 12 Chestnut Street. The disease eventually killed ten doctors and Hiltzheimer, but not Benjamin. He volunteered to serve at Bush Hill outside the city where many of the city's sick were sent to recover or die. He labored there under the scholarly French doctor, Jean Devèze, treating 807 recorded cases.

Edward's sister Elizabeth, widow of Dr. Samuel Swift, died in October 1795 at age 78, perhaps from yellow fever which was again raging. In 1796 Edward's wife, Catherine, died on October 6. A single published obituary stated: "On Thursday the 6th instant, departed this life, in the sixty-ninth year of her age, Mrs. Catherine Duffield, consort of Mr. Edward Duffield, of Benfield, in the County of Philadelphia. They had been united near half a century in the bands of matrimony."[31]

More sad news came in February, 1797, with the death of Edward's daughter-in-law: "On Saturday last, departed this life in the 43d of her age, of a sudden, short, but excruciating illness; Mrs. Rebecca Duffield, wife of Dr. B. Duffield of this city."[32] In the collection of Pottsgrove Manor is a handwritten copy of a letter from the widower to his sister-in-law, Miss Ruth Potts:

Saturday evening Feb.4th, 1797.
It is extremely easy in the serene moments of health to talk of affliction in the stoic indifference, and boast of an impenetrable apathy to feeling. But when the keen stab is made that robs us of a parent, a child or a wife; how feeble is the exaltation of firmness, and how prostrate does our vaunted resolution lie. I have felt sorrow it is true — I have in my life had its share of bitterness, but my dear Ruth until this day, this black, this dreary, dark, and never to be forgotten day, I never knew what the exacerbation of sorrow was. Your sister this day left me forever. I write by the side of the bed in which she expired, while she lies lifeless and cold almost in my sight, the daughter of her name, blest innocent, waits for my arms in her mothers bed to embrace me, and wipe her tears on my bosom. I can no longer collect myself, I can write no more.
Farewell, B. Duffield."[33]

We do not know if Benjamin wrote similar lines to his father, but Supreme Court Justice James Iredell, a close friend of Benjamin's, informed his wife of the passing of "our excellent friend Mrs. Duffield."[34] In 1793, Judge Iredell received a letter from North Carolina Governor Samuel Johnston (1733–1816) indicating that he, too, was close with both Benjamin and Edward: "I am much pleased to hear of Dr. Duffield's success. I have a great partiality for him; and indeed, for all of his father's family with whom I was acquainted, as much for the old gentleman as for himself, &c."[35]

Benjamin predeceased his father on December 15, 1799, and was buried at Christ Church. Age forty-six, he was weakened by yellow fever and by his unending exertions for others. He no doubt was dispirited, as well, by the loss of his wife. His two sons did not follow in the medical profession, but his three daughters married physicians he trained.

**Figure 1.5** Edward Duffield's grave, tabletop at center, at All Saints' Torresdale Episcopal Church, Philadelphia. Author photo.

Edward Jr. and John Church (but not Edward) administered Dr. Benjamin's estate. They advertised on January 22, 1800, that those who owed and were owed money should furnish their accounts at 266 South Second Street.[36] Shortly thereafter, understandably affected by the recent deaths in his immediate family, Edward wrote his will, dated July 28, 1801.

In 1802, on November 6, Edward's widowed sister, Sarah Kinnersley, died at age eighty-one, and Edward followed her on July 12, 1803. He was interred alongside his wife at All Saints'. His passing was inscribed in Quaker meeting records, indicating his long friendships with Friends in the area, but not announced in any known newspaper notices or obituaries.

Today his weathered table-style gravestone (Figure 1.5) stands in Torresdale alongside monuments to several family members including his two daughters, Elizabeth Ingraham (1761–1851) and Sarah Hepburn (1755–1837), his son Edward Jr. (1764–1836), and officers who fought on both sides of the Civil War. Other daughters who died young were buried at Christ Church with Edward's grandfather.

Edward specified the text carved onto his stone, without mention of clockmaking: "Under this stone rest the remains of Edward and Catherine Duffield, late of the Manor of Moreland. Edward Duffield was born April 30, 1730 and died July 12, 1803 Aged 73 years 2 months 12 days. Catherine Duffield was born October 11, 1727, Died October 6, 1796 aged 68 years 11 months 25 days."

Edward apparently lived well during his final years at Benfield, as expected from his wealth and status. His multi-page estate inventory noted many fine furnishings and household items as well as seven decanters, 192 gallons of beer, thirty bottles of sherry, four demijohns of whiskey, two other cases of liquor, seventeen bottles of wine, and 453 dollars in gold.

# POSTSCRIPT

Edward Jr. lived unmarried at Benfield until his death in 1836. His share in the Library Company of Philadelphia, along with his violin, went to his great-nephew Benjamin Duffield. Several clocks and watches, without their makers' names noted, were bequeathed to Alfred and Edward Ingraham, Martha Neill, and Rebecca Martin. The Benfield estate was sold to John and Patrick Murray, and today its extensive acreage in three townships—Moreland, Byberry, and Lower Dublin—includes Northeast Philadelphia Airport, the Torresdale Frankford Country Club, and many housing developments.

Online searches for "Edward Duffield" bring multiple entries for Edward's great-grandson, Edward Duffield Neill (1823–1893) (Figure 1.6), mentioned above. A direct descendant and well-published historian, he documented much of what we know about

**Figure 1.6**  Edward Duffield Neill 1849 daguerreotype, Courtesy of the Minnesota Historical Society.

the Duffield family in his 1875 privately-printed book, *John Neill of Lewes, Delaware, and His Descendants*. The Reverend Neill, grandson of Edward's son Benjamin, was born in Philadelphia and was educated at the University of Pennsylvania, at Amherst College, and at the theological seminary in Andover, Massachusetts (the historic town where this book's author resides). He then lived and worked as a Presbyterian minister, missionary, and educator in frontier Minnesota.

In 1864 Neill took charge of President Lincoln's correspondence and remained at work in the White House until 1867, also serving under President Andrew Johnson. He returned to St. Paul and was founder and first president of Macalester College in 1872, but his name was removed in 2019 from a main school building. Neill's offences included desecrating Native American graves during archeological digs, describing those indigenous people as savages, and refusing to allow women students into his classes. Some faculty and administrators, however, have objected to this "presentism" attack on past attitudes and practices.

References to his many publications also fill the online search pages. At the Boston Athenaeum, where this author is a Proprietor, a search of its catalogue resulted in fourteen titles on the shelves. (The family history cited above is not among them.) All relate to some aspect of early American history on which Reverend Neill was expert. The titles include:

*The History of Minnesota: From the Earliest French Exploration to the Present Time* (1858)

*Concise History of the State of Minnesota* (1887)
*The English Colonization of America during the Seventeenth Century* (1871)
*Virginia Company of London: Extracts from their Manuscript Transactions* (1868)
*The Fairfaxes of England and America in the Seventeenth and Eighteenth Centuries* (1868)
*History of the Virginia Company of London* (1869)
*The Founders of Maryland as Portrayed in Manuscripts* (1876)
*Terra Mariae, or, Threads of Maryland Colonial History* (1867)
*Virginia Carolorum: The colony Under the rule of Charles the First and Second, A.D. 1625 – A.D. 1685* (1886)
*Virginia Vetusta: During the Reign of James the First* (1885)
On the lighter side was a 25-page booklet: *Pocahontas and Her Companions* (1869).[37]

The Government Printing Office published his 1868 letter to Hon. N.G. Taylor, Commissioner of Indian Affairs: "Effort and Failure to Civilize the Aborigines."[39] Neill provided a detailed overview of interactions between European colonists and Native Americans, describing policies to civilize, Christianize, educate, and assimilate the continent's original inhabitants. Neill explained why these efforts mostly proved unsuccessful. A biography of Reverend Neill by Huntley Dupre was published in 1949 by Macalester College Press.

An even more notorious descendant was the son of Edward's daughter Elizabeth. Edward Duffield Ingraham (1793–1854) was a Philadelphia lawyer who served as United States Commissioner enforcing the Fugitive Slave Law. He personally approved the seizure and remanding of many escaped freedom-seekers. The *Rhode Island Freeman* celebrated his sudden death by apoplexy in a November 12, 1854 dispatch:

> The great event, since my last, is the death of slave hunter Edward D. Ingraham. He was the United States hunter, you know, being the Fugitive Slave Law Commissioner.... Cut off with all his imperfections on his head, he had a fearful account to settle—somewhere. As usual, our papers united in eulogising one of the most detestable men that ever insulted the venerable Hall of Independence by his presence.

The *Philadelphia Bulletin* was kinder. It recalled the high esteem he earned as a member of the Philadelphia bar, and added:

> His professional and official duties, however, by no means occupied all his time. There are few among us whose reading in all the branches of the lighter literature of England and this country has been so extensive; none whose research into matters of historical interest has been more industrious ... His collection of books, autographs and antiquities, though miscellaneous, confused, and bearing many markers of the eccentricities of the collector, are probably equal in value to those of any other person in the country.[39]

He published a short book in 1849: *A sketch of the events which preceded the capture of Washington by the British, on the twenty-fourth of August, 1814.*[40]

His collections were sold in 5,158 lots on March 20, 1855, at an American art auction. As described in Chapter Five, one of just two known letters written to Edward was discovered at the Library of Congress in a book formerly owned by this grandson. This book perhaps was one of those auction lots. If E.D. Ingraham had more materials related to his clockmaker grandfather, they remain undiscovered.

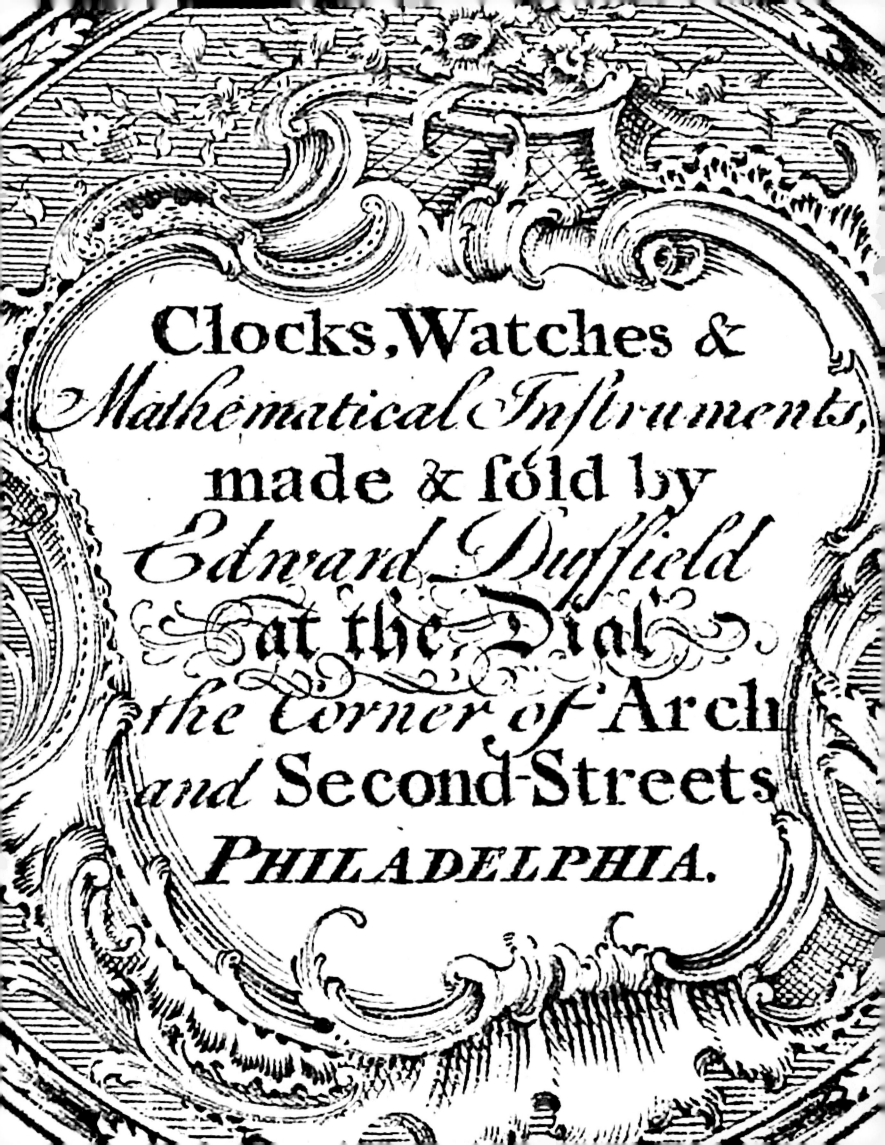

CHAPTER TWO

# Edward Duffield: Merchant, Engraver, Clockmaker and Watchmaker

*Probably more craftsmen achieved social distinction at Philadelphia than in any other community, for there the crafts reached their highest development and prosperity and more artisans amassed wealth.*

— Carl Bridenbaugh, *The Colonial Craftsman*, 1950, p.157

In April of 1751, Edward reached age twenty-one and opened his food shop at one of Philadelphia's busiest crossroads at the northwest corner of Second and Arch streets, a property he inherited. He lived and worked there until 1775. Diagonally opposite was the bustling George Inn (see Figure 7.3).

His first newspaper advertisement for this venture appeared August 1, followed by another on the 8th and the 15th (Figure 2.1).

Sometime thereafter, he changed his business from stocking imported delicacies to selling and servicing clocks, watches and other instruments of measurement.

We have an image but no original of an undated watch paper that Edward, like many of his colleagues, pressed into the backs of pocket-watch cases (Figure 2.2). On the paper, he called himself a maker and seller and implied that there was a "Dial" sign or exterior clock at his shop's location.

We do not know when Edward had these little paper disks printed and officially became a horologist, instrument maker, and engraver. No known ledgers or account books document

*Detail of Edward Duffield watch paper (see Figure 2.2).*

**Figure 2.1**  Newspaper advertisement placed by Edward Duffield, *Pennsylvania Gazette*, August 15, 1751, p.3. Courtesy of Newspapers.com

his business. No newspaper published his vocation until an advertisement five years after his shop opened when as "Edward Duffield, Watchmaker" he alerted the public to a silver spoon, likely stolen, that he would return to its rightful owner (Figure 2.3).[41]

We can, however, imagine a likely path for his transformation from aspiring city merchant to watch- and clockmaker. Edward was fortunate in that no trade or guild restrictions in colonial Philadelphia obstructed his path. From 1751 until he was first identified as a watchmaker five years later in 1756, he had ample time to become skilled at a craft that had perhaps begun as a hobby. He could have learned from experts close at hand and by practicing diligently on his own. Another six years passed, providing even more time to gain expertise, before he began maintaining the city's State House clock in 1762. He succeeded the clock's maker, Thomas Stretch (1697–1765), who with his father, Peter, could have been among Edward's vocational guides. For private-customer repairs and sales, his family, civic, and church connections would have gained him ready access to the city's upper crust. This may mirror the background of his more-famous contemporary, David Rittenhouse, whose family was well-off and whose interest in horology came early in his life.

**Figure 2.2** Undated Edward Duffield watch paper. Courtesy of Winterthur Museum, Garden & Library: Joseph Downs Collection of Manuscripts and Printed Ephemera, Doc.499, p.5.

At the risk of imposing contemporary hindsight, I can speculate that Edward's path echoes my own road to horology more than two centuries later. Age twenty-eight in 1980 with a keen interest in mechanical tinkering, I developed a love for horology that grew until it bloomed into a full-time profession a dozen years later. I was mentored but not apprenticed. I became adept mainly through study, practice, friendships, and experience. My comfortable upbringing, college education, and white-collar jobs eased financial pressures and nurtured confident relations with my customers.

Edward at no time publicly advertised his horological services and products, with a single exception in 1770 when he offered for sale a "Genteel Eight Day Chamber Clock"[42] that he did not claim to have made. Edward was not alone in this lack of self-promotion. David Rittenhouse never advertised, nor did Peter Stretch, Edward's possible teacher, yet these men were professionally successful. Many grand clocks by Rittenhouse and the Stretches were and are as prized as Edward's. Yet, as summarized and transcribed in Chapter Four, many other Philadelphia clockmakers and watchmakers regularly advertised repairing and selling of clocks and watches. John Wood, Sr., for example, frequently placed ads and was an active merchant. His imposing tall

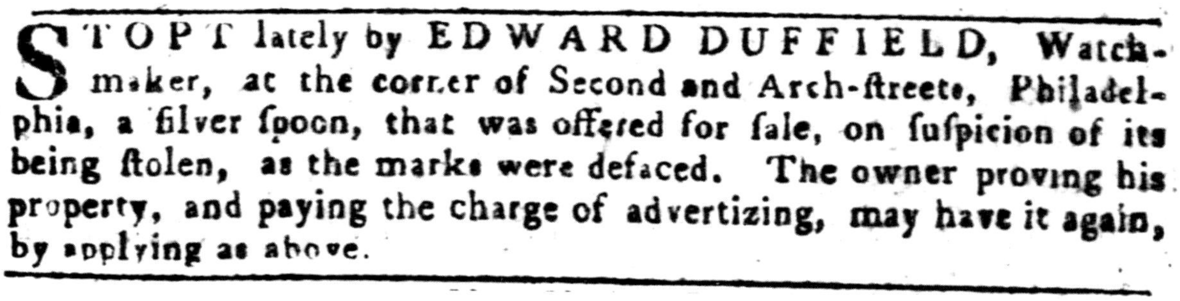

STOPT lately by EDWARD DUFFIELD, Watchmaker, at the corner of Second and Arch-streets, Philadelphia, a silver spoon, that was offered for sale, on suspicion of its being stolen, as the marks were defaced. The owner proving his property, and paying the charge of advertising, may have it again, by applying as above.

**Figure 2.3** Ad placed by Edward Duffield, *Pennsylvania Gazette*, July 8, 1756, p.4. Courtesy of Newspapers.com.

clocks grace museum galleries and private collections today. This suggests that Edward was in a separate and higher class of elite, skilled artisans worthy of our attention and study. He had no need to self-promote. He served only the city's richest citizens who knew him and knew what he did. He was one of those richest citizens himself and he was not desperate for patronage. He was a gentleman, but one willing to work with his hands. His clocks were expensive, and Edward sold relatively few of them.

The following is what we know and can surmise about Edward's work as a skilled artisan. His work included not only the making, selling and repairing of clocks and watches, but also engraving and the construction of surveying instruments.

Edward certainly sold watches and clocks. The paper receipt for a new silver watch purchased in 1767 by Joseph Shippen is at the Historical Society of Pennsylvania (Figure 2.4).[43] Shippen paid £14 for the watch and £2.6 for a glass, presumably a spare since the thin crystals covering the dial and hands were easily broken and were frequent repair items.

Winterthur's archives hold a small account book belonging to Philadelphia cabinetmaker Benjamin Randolph. An unspecified purchase from Edward was recorded at the bottom of one page, written and signed in his hand (Figure 2.5): "Rec'd Dec. 24 [... possibly hereby] of Mr. Benj. Randolph one pound 1/ in full £1.1. Edw. Duffield."

London printer David Strahan wrote in 1768 to Philadelphia printer David Hall, Franklin's former partner, about Edward selling watches for him: "You have done extremely right to put the Watches into such proper Hands. I know that they are cheap and am therefore hopeful they will be disposed of by Mr. Duffield in a moderate Space of Time, without much (if any) loss."[44]

Franklin likely encouraged his longtime correspondent and London friend to seek Edward's help (Figure 2.6).

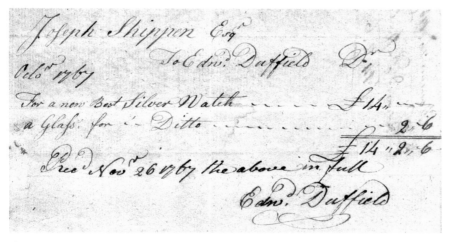

**Figure 2.4** October, 1767, bill and receipt for Joseph Shippen, Collection of the Historical Society of Pennsylvania, *Etting Papers*, Vol. 1, Folder.123.

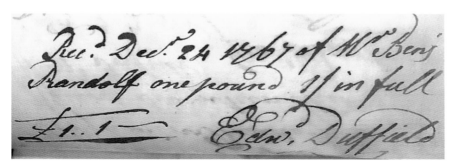

**Figure 2.5** Benjamin Randolph receipt book entry, collection 337, 54x549: *Benjamin Randolph Financial Records, 1763–1783*, p.106. Courtesy of Winterthur Museum, Garden & Library: Joseph Downs Collection of Manuscripts and Printed Ephemera.

In 1770 Edward placed his sole newspaper ad for a clock he was selling. It was not of his own making since smaller spring-driven clocks by him are unknown. It may have been the English-made clock, with his name on it, now at Winterthur—Catalogue No.6—or another he imported or repurchased from a customer.

Edward made many of the clocks he sold, but while he and his competitors also identified themselves as watchmakers, this does not signify that they actually made watches, as sometimes was asserted at the time and claimed today. The term "watchmaker" then and now describes a skilled craftsman who cleans and repairs watches, rarely somebody making each tiny gear, spring and screw. Edward and his fellow Colonial watchmakers, like their English counterparts, could have assembled and finished watch movements from pre-made rough parts. The 1747 excerpt heading Chapter

A GENTEEL Eight Day CHAMBER CLOCK, almost new, which Chimes the Quarters, to be sold, by EDWARD DUFFIELD, at the Corner of Arch and Second-Streets.
N. B. Wanted in two Weeks, a sober industrious man, who understands driving a Team, and taking Care of Horses. Enquire of said Duffield.

**Figure 2.6**  Advertisement placed by Edward Duffield selling a chamber (or table) clock, *Pennsylvania Gazette,* November 8, 1770, page 4. Courtesy of American Antiquarian Society, accessed via GenealogyBank.

Three confirmed this, as does Appendix 4 listing forty-three "artificers" employed in making a watch. Incidentally, the first documented hand-made American watch may have been by John Cairns (1751–1809) of Providence, Rhode Island[45], in around 1800. He advertised: "Plain watches made, of any fashion required, for 25 Dollars: Horizontal 28, warranted for two Years without Expence, except in case of Accident."[46]

George Eckhardt added:

In these early days there were men in Philadelphia who sold watches, but the movements were usually imported from England, and later from France. Even if the watch carried the signature of a Philadelphia watchmaker, the parts were imported…. For instance, there is an extant watch which Edward Duffield "made" for Franklin. The value of this watch lies largely in the fact that it belonged to Franklin rather than that it was "made" by Duffield.[47]

The whereabouts of this Franklin watch are unknown today. Another Franklin watch, however, gifted to him by his close English friend David Hartley upon the signing of the 1783 treaty ending the American Revolution, is owned by Temple University.[48] Edward may have admired this one when his friend returned home from Europe.

In 1769 the APS commissioned Edward to construct a precise timepiece to observe the November 9 transit of Mercury, although a Rittenhouse clock used in the recent transit of Venus perhaps could have been requisitioned. He completed the job within three weeks. The term "timepiece" applies to a non-striking clock that would have greater accuracy. A committee approved Edward's work and authorized payment of £15.17.6 in the following February. This timepiece, No.3 in the Catalogue, remains in the APS collection.[49] A subsequent report by Thomas Bond and local scientist, John Ewing, stated:

a new Time-Piece made by Mr. Duffield of this city, with an ingenious contrivance of his, in the construction of the pendulum, to remedy the irregularities arising from heat and cold…[50]

This author examined the movement and pendulum and saw no such contrivance. As confirmed previously by Murphy Smith,[51] the simple eight-pound pendulum is not compensated for temperature changes. Possibly a unique Duffield-crafted compensating pendulum was replaced and is lost to Duffield scholars. There is an unusual pendulum-suspension device for fine slow-fast adjustments, but that had no role in temperature compensation (Figure 2.7). In 1770, an Edward-signed pocket watch—No.12—was lost and a thirty-shillings reward was offered for its return (Figure 2.8).[52] No.7 was reported lost on another occasion. Another, with no number noted, was repaired three times for Mr. William Swet by Boston watchmaker Richard Cranch. This could indicate that at least twelve Duffield watches existed although none are known today. These almost certainly were English watches on which Edward engraved his name.

A Duffield clock was in Newton Township (Gloucester County, New Jersey) in 1771 when its owner, Elizabeth Miekle, offered a £3 reward for its return: "feloniously taken, stolen, and carried out of the Dwelling-House… one Eight-Day clock, Arch-faced, with a rais'd circular Plate in the

**Figure 2.7** Side view of Duffield movement, Catalogue No.3, showing pendulum-suspension system allowing fine slow-fast adjustments. Note extra-thick wood seatboard for additional strength and stability. John Wynn photo.

arch, and the maker's name EDWARD DUFFIELD, engraved on said Plate: The clock was broken off from the case, and the Weights and Key left."[53] There is no record of the clock's recovery and reunion with the parts left behind, but it may be one of the clocks illustrated in the Catalogue.

Edward's ingenuity in constructing clocks was reported in a favorite anecdote about Edward originating in the 1881 *Annals* by John Fanning Watson "who was not always sufficiently critical of the sources of his information."[54] Watson wrote more than a century later about a clock that Edward mounted outside his shop:

It was so rare to find watches in common use that it was quite an annoyance at the watchmakers to be so repeatedly called on by street passengers for the hour of the day. Mr. Duffield, therefore, first set up an outdoor clock to give the time of the day to people in the street.[55]

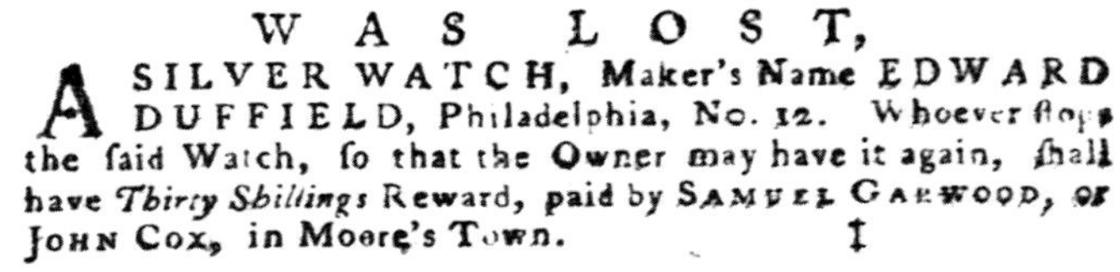

**Figure 2.8** Advertisement placed by John Cox seeking lost watch signed by Edward Duffield, *Pennsylvania Gazette,* June 14, 1770, p.5. Courtesy of Newspapers.com.

The *Annals* two-paragraph biography of Edward enhanced the tale:

> When he kept his shop at the north-west corner of Second and Arch streets, he used to be so annoyed by frequent applications of passing persons to inquire the time of day—for in early days the gentry only carried watches—that he hit upon the expedient of making a clock with a double face, so as to show north and south at once, and projecting this out from the second story, it became the first standard of the town.[56]

Finally, a 1911 history of Lower Dublin Township repeated and enlarged this story but provided no technical details. George Washington was added, and Ben Franklin was given credit for the idea although he is not known to have written about it. Edward's converting the clock to a single dial when he installed it at the Lower Dublin Academy is better verified, but that 1911 memoir was drafted more than a century later:

> On October 23, 1802, the Board presented a vote of thanks to the President, Edward Duffield, for his indefatigable industry to the Building Committee and for his great kindness in presenting a clock to the institution, which was set in motion on the 20th instant. The clock deserves more than a passing mention, for its donor, Mr. Duffield, was a man of much consequence in the early annals of Philadelphia. He was a clock maker, and his shop was the resort of all the well-known men. He was Franklin's executor, and Washington frequently stopped to ask the time, no doubt a welcome visitor; but when the vulgar crowd claimed the same privilege, Benjamin Franklin disliking to see his friend so frequently interrupted, suggested that he make a clock with two faces, placed out obliquely from the store at Second and Arch Street, so that it could be seen by persons approaching from either direction. The clock was a great novelty for that period and became the standard timekeeper for the town. This was the one that Mr. Duffield gave to the Academy. It had an old cannon ball, weighing 30 pounds, for a weight, and remained in use for many years. It was placed in a dormer window opened especially for it, with a clock dial painted white with large black figures. After getting out of repair, it was laid away in the attic, but in 1875 was put in order and again used. When the property was sold the Trustees retained it, but its usefulness as a timekeeper is gone. It should, however, be retained as an historical relic of much value.[57]

An early indistinct photograph hints at the clock in place at the school. A recent photo (Figure 2.9) shows the large opening atop the building originally designed

**Figure 2.9** March, 2020, photograph of former Lower Dublin Academy. Author photo.

by Edward, as reported by Edward Duffield Neill who wrote: "Edward Duffield was instructed to prepare the plans of a new building. When erected, it was fifty feet long, thirty feet wide and two stories high."[58] The clock itself disappeared long ago.

Public clocks providing local time in Philadelphia were noted before Edward's, however. Scottish tourist Alexander Hamilton saw one in 1744 and Thomas Stretch completed the large State House clock in 1753, two years after Edward began selling lemons. As well, while two-sided public clocks were hanging or projecting from walls by late in the next century, especially in railroad stations, almost no examples are known from before then, so the story may well be apocryphal. When Edward's clock later was adapted and installed in a dormer opening at the Lower Dublin Academy, the dial must have been at least three feet across. Yet this prominent and unique outdoor clock, located at one of Philadelphia's busiest crossroads, was never noted by period diarists, newspaper reports, artists, visitors, or in Joseph Fox's detailed 1758 fire-insurance survey.

A clock affixed to the outside of his building would have required a complicated method of driving the two pairs of hands. Unlike later spring-powered or electric versions with movements situated between the twin dials, Edward's machine was weight-driven and could not have been made in that way. As in similar installations at other clockmaker shops with exterior clock dials, the large movement would have been indoors and upstairs, safe from the elements, with horizontal drive shafts through the exterior wall. Additional gearing would have been needed to rotate the hands in opposite directions on each face.

Typically, as mentioned, Edward, like most watchmakers of his time, would have been much busier repairing and maintaining clocks and watches than in making new ones. Ticking timepieces needed frequent service due to dusty, smoky, and damp environments, worn and broken parts, user abuse, jolts on horseback and wagons, and drops during viewing and winding. He could have charged, as many clockmakers did, a fee for winding, setting, and adjusting customers' clocks during regular house calls. He would have arrived with his tools, oils, and his accurately set watch to bring the correct time.

In 1761 and again in 1764 Edward repaired a watch for political leader James Hamilton.[59] In 1762 he charged Stephen Collins six shillings for installing a new clock bell and he repaired the merchant's watches in 1763 and 1772.[60] In 1762, Edward advertised a generous twenty-shilling reward for the return of a stolen silver watch signed by London maker William Grant,[61] perhaps a watch he had sold or serviced.

Edward was appointed in 1762 to care for the State House clock, taking over from Thomas Stretch, earning £20 per year for this prestigious public contract.

No detailed description of Stretch's clock exists, but what is likely to be a similar wrought-iron movement is illustrated in Chapter 4 (Figure 4.3). We know that its dials were at opposite ends of the long building and that the movement was mounted at the center of the building, requiring lengthy connecting rods between the movement and its two pairs of hands. This heavy, high-friction arrangement needed special attention in addition to routine lubrication and adjustment. Maintaining this public timepiece likely was stressful since everyone would know if the clock malfunctioned.

In 1763, the Assembly referred to committee Edward's bill for care of the clock, and then later voted to pay him £76.14.1. This indicates either that the clock required more attention than was covered by the £20 annual stipend, or that he was being compensated for unrelated work as well. Edward waited three years to be paid and finally received £61.1.6 in public-debt certificates, not cash. In the "Estimate of the Debts Due from the Province of Pennsylvania, 1769" Edward was owed £20 for the year's maintenance of the State House clock. The Assembly voted to pay him £20 and approved an additional £40.9.0, again for extra work. In 1772, the Provincial Council approved

**Figures 2.10 & 2.11**  1757 Indian peace medal, front and back. Author photos.

another £23 for Edward's care of the State House clock and paid him £20.15.0 for past services. In the same year, the Council made much larger payments to David Rittenhouse for instruments he provided to study the flows of the Susquehanna and Schuylkill Rivers. When Edward moved to his country estate in 1775, he announced that he would discontinue care of the State House clock, leading David Rittenhouse to propose to the Assembly:

> as he the Petitioner has the Care of the Time-Piece, belonging to the Philosophical Society, which is kept at the Observatory in the State-House Square, with astronomical Instruments for adjusting it, he conceived it would not be inconvenient for him to take Charge of the said public Clock; and therefore prays the House will be pleased to appoint him to that Service, when Mr. Duffield shall discontinue the same.[62]

The time accurately determined by Rittenhouse at his nearby observatory may already have been consulted by Edward to regulate the city's public timekeeper. After Edward's relocation, the Provincial Council voted in September to pay £10 for his previous six months of State House clock maintenance, and to pay David Rittenhouse the same amount for the next half year. Edward also was paid for repairing a pump.

Edward fabricated instruments of measurement used in surveying. Duffield-signed surveyor's compasses are known and are included in the Catalogue. Further evidence of Edward's instrument work is in the archives of the APS. *Minutes and Papers of the Mason and Dixon Survey, 1760–1768* show two entries in the 1761 expenses ledger indicating that Edward was paid for work prior to the famous surveyors' landing in 1763. The first on April 28: "To cash paid Edward Duffield for making a measur.g Instrument, 2 Spirit Levels for Do, Iron & woodwork for ditto as p.Voucher No.25. 6.16." The second was in June: "To cash pd. Ed. Duffield for repairing Instruments as p. Voucher No.71. 3.13.0."

An anecdote, if true,[63] also provides support for this. In 1755, young Edward may have accompanied General Edward Braddock's doomed massive offensive into Western Pennsylvania during the French and Indian War. Local historian Joseph C. Martindale reported that Edward "entered Braddock's army as an assistant surveyor,"[64] and that Edward was present when a surveying instrument was being misused by a British officer and at one point Edward commented upon this error.

When Edward's expertise was questioned, Edward replied that he had made the instrument.

General Braddock's Orderly Book, Library of Congress, p.XV, reports: "One corporal and eight men of the Line to attend at 6 O'clock every morning, to assist the Engineers in Surveying." One of these men may have been Edward, who may have joined them at the start of the expedition, when a small portion of Braddock's army marched to Pennypacker's Mills near Philadelphia to join contingents coming from Virginia and Maryland. If he did subsequently witness the debacle and barely avoided being killed, as did twenty-three-year-old Lieutenant-Colonel George Washington, this may account for his later paying a fine to avoid military service in the Revolution. However, Edward's presence there is not confirmed in any existing records and there were no Pennsylvania troops involved, only teamsters recruited by Franklin at his own expense. Franklin's brother-in-law, John Read, was one of those wagon masters, however. He was acquainted with his one-time neighbor, Edward, and may have brought him along.

Oddly, a 1757 letter maligned Edward's expertise in this field. An unidentified Philadelphia clockmaker requested that Franklin in London deliver a proposal for a new method of determining longitude at sea. The writer, hoping for a reward, was not personally acquainted with Franklin. Within the document, wrongly addressed to the Royal Society rather than the Board of Longitude, he attacked "ED," almost certainly Edward since no others had those initials:

> I Must Acknowledge he is a better Nitting Nedle Case Maker than a Clock Watch or mathematical instrument Maker Witness Governor Tinker Reflecting telescope Which I made a New Mirror and alterd the focus of the Glasses he Spoild.[65]

John Tinker (1700–1758) was a British governor of the Bahamas who came to Philadelphia in 1755 to inaugurate a new Masonic lodge. A 1790 auction conducted by John Patton, however, included a more positive but equally rare mention of Edward's instrument work: "… belonging to the estate for the late Surveyor General, John Lukens, Esq…Theodolite… by Cole of London, improved by Duffield…"[66]

Edward also did engraving. In 1756 he produced dies for the first two commemorative medals struck in colonial Pennsylvania. To do this, he cut the text and designs onto the flat faces of metal punches that fit into sockets. Both sides of each medal were struck by another craftsman with blows of a sledgehammer. Only limited quantities of medals could be produced because the dies became worn, dull, or broken.

Edward's first assignment was a peace medal commissioned by Quakers (Figures 2.10 & 2.11). They lost control of the colony's government and were encouraging non-violence through their Friendly Association for Regaining and Preserving Peace with the Indians by Pacific Measures. The medals, showing a bust of King George II on one side and a peace pipe being shared over a council fire by a Quaker and Native American on the other, were presented to Native American leaders in hopes of avoiding further bloodshed. Edward was paid £15 for this engraving work.

These were the first such medals minted in the Pennsylvania colony, precursors to many peace medals

**Figure 2.12** Henry Inman, 1832–1833, Pes-Ke-Le-Cha-Co Pawnee Chief. Courtesy of Metropolitan Museum of Art, 2018.501.2.

**Figure 2.13** Kittanning Destroyed Medal, courtesy of Yale University Art Gallery, transfer from the Yale University Library, Numismatic Collection, 201, Bequest of Charles Wyllys Betts, B.A. 1867, M.A. 1871.

distributed to Native American chiefs. However, they were not the first medals struck in America. A much-earlier copper-alloy example[67] made in 1676 is on display at the National Museum of the American Indian in lower Manhattan. Most likely it was presented by the Massachusetts Bay Colony to one of the eighty Indian warriors recruited into a company led by Captain Samuel Hunting during King Philip's War. Many period portraits show such peace medals proudly displayed on the chests of Native American leaders (Figure 2.12).

Conversely, his second medal, commissioned by the Corporation of the City of Philadelphia, celebrated the September destruction of the Delaware Indian village at Kittanning near Pittsburgh (Figure 2.13). This was the first medal struck in British North America to honor military exploits. One side showed the village being burned; the city's corporation arms were on the obverse. During the attack, colonial militiamen slaughtered Native Americans while suffering forty-nine casualties themselves. Militia leader Lieutenant-Colonel John Armstrong of Carlisle, who was wounded in the attack, and several of his subordinates, were awarded these medals by Mayor Atwood Shute who praised their courage and bravery. Armstrong's sister, Margaret, was the second wife of Edward's Presbyterian cousin, Reverend George Duffield (1732–1790).

Edward is known to have made a single sundial, which was dated 1757 and is No.71 in the Catalogue. It has an uncommon finely-engraved scale of minutes to add or subtract when converting to clock time. Since a sundial's readings vary seasonally by up to sixteen minutes, this extra feature would have been useful for owners of clocks and watches wanting to set them accurately from the sundial, and it is a fine example of Edward's expertise. Edward may have made it specifically for himself or for a customer. Owners of sundials without this built-in day-by-day conversion table could consult annual almanacs, including Franklin's, that routinely included pages listing this information.

Some writers have claimed that carpentry tools in Edward's estate inventory suggest that he built cases. That would have been extremely uncommon. Nothing indicates that Edward had training, skills, and experience with cabinetmaking.[68] The woodworking tools listed in Edward's estate had other practical uses. We know, for example, that a few years prior to his death, his estate was producing large quantities of wooden barrels. His 1803 inventory listed quantities of oak, white pine, poplar, and "sap" planks, but significantly no costly walnut and mahogany lumber appropriate for clock cases. While some clockmakers in rural areas produced their own cases, having no nearby competent joiners, they made simpler cases in cheaper woods for down-market local customers, not like the elegant cases containing Edward's clocks. Edward had several competent cabinetmakers within shouting distance.

In 1775, Edward decided to sell his business and retire to his country estate, and there is little or no evidence that he continued his clockmaking. While broad date ranges extending into the 1790s have been cited for the years he produced clocks, it seems clear that he retired from the trade when he departed Philadelphia in 1775 at age forty-five. Tax records as early as 1779 list him as a farmer not a clockmaker, a 1793 document identifies him as the 'late' clock and watch maker, and the vocation was not inscribed on his gravestone.

Perhaps like his artisan friends, printer Ben Franklin and cabinetmaker Jonathan Gostelowe who each in their forties gave up manual labor, Edward was ready to leave the workbench at about the same age. Perhaps, too, failing eyesight or physical ailments were a factor. But just as likely were a few other realities: the growing conflict with Britain assuredly affected both the market for his services and the availability of English supplies, and Edward may have realized that flaring tensions with England could eliminate demand for his clocks. His decision could also have been spurred by a growing number of competitors—Philadelphia's 1774 tax list indicated that 934, thirty percent, of the city's 3,432 property owners were craftsmen of some kind.[69] Several of these were active clockmakers and watchmakers. There may have been the lure of healthier, quieter living on his country estate. Finally, his wife Catherine may have had a say in the matter as well.

At the time of his move, Edward leased his shop to clockmaker John (formerly Johannes) Lind. He may have sold Lind some tools and equipment that later were absent from his estate inventory. Lind advertised in April and twice again in May:

> JOHN LIND, WATCH AND CLOCK-MAKER, Begs leave to inform the Public, and his Friends in particular, that he has opened shop at the corner of Arch and Second-streets, in the house lately occupied by Mr. Edward Duffield, Watch-maker, where he carries on the said business in all its various branches, and at the most reasonable rates. He thinks it unnecessary to enlarge on particulars, as his abilities are sufficiently known in this city. WANTED, an apprentice to said business.[70]

After moving permanently to Benfield, Edward continued to "devise machines and apparatus... and invented a horse-rake and an odometer,"[71] but most likely his time was consumed running his large farm and pursuing gentlemanly and civic activities.

An estate inventory made as part of an estate appraisal on September 29, 1803, by Joshua Comly, Derrick Peterson, and Jacob Sommer, shortly after Edward's death,[72] did list clocks (movements without cases), but these might not have been made recently or made by him, especially those listed as unfinished musical and alarm. There are no signed Duffield examples of these two styles. Musical clocks are quite complicated, costly, rare, and require several additional skills to construct. The alarm named in the inventory was most likely a smaller wall-hanging clock with an alarm function, perhaps an early English lantern clock or a wag-on-the-wall of German or Pennsylvania-German origin.

Some scholars, dealers, and curators suggest that the tools, brass sheets and wire, and related supplies included in the inventory and listed below are signs that he still worked at the bench. However, these too could have been leftovers, like the movements, and used instead for many other purposes around his properties.

The following items have been extracted from the estate inventory because of their horological relevance. The first item would be a non-striking floor-standing clock, given its "Time Piece" name and high value, but it cannot be correlated with any Duffield clock known today. The "jack" on the fifth line most likely was a roasting jack described in Chapter Four as clockwork cooking devices sometimes made by clockmakers, although none by Edward are known.

| | |
|---|---|
| 1 Time Piece Walnut Case | 45 |
| 1 Small case cont Instruments | 3 |
| 1 Shagreen case watch | 8 |
| Tin Scales & Weights | 2.50 |
| 1 Jack  1 Spit | 4 |
| 1 Watch Case | 10 |
| Gold 56dwt 68  a 6/ p dwt | 453 |
| Large turning wheel | 2 |
| Books (some possibly horological) | 85 |
| 14 1/2 lb. Whale Teeth | 6.44 |
| 4 lb 11 oz Turtle shell | 14.6 |
| Lot of seals & old springs | 12 |
| Lot of brass wire plate etc. | 13 |
| 50 lbs sheet brass @ 2/4+ 4 pieces $25/100 | 15.81 |
| 143 1/2 Old  Do  @14 d | 22.91 |
| 3 oz  2 danti9?) silver bows and studs | 2.10 |
| Screw plate & Tools | 1.20 |
| 2 Boxes with Drawers and Lot of Steel | 1.50 |
| Ivory watch tools and Vices | 2.25 |
| Seal Files & punches | 4 |
| Brass Steel Turtle Shell etc. | 1.92 |
| 2 Clocks Compleat @$16 | 32 |
| 3 Do unfinished | 42 |
| 1 Do Musical Do | 8 |
| 1 alarm | 10 |
| Lot of Ivory, Wire etc. | ? |
| 1 Lot Watch Chrystals, Materials, …. | 10.90 |

Several inventories exist for clockmakers of Edward's time who remained at the bench until their deaths. In the inventory of Connecticut clockmakers who were contemporary with Edward, John Avery's (1732–1794) most valuable item was the "brass lathe" valued at £18.0,[73] a "clock engine" of Joseph Carpenter (1747–1804) was worth $16.6,[74] and that state's most famous maker, Thomas Harland (1735–1807), left a fifteen-dollar "clock engine."[75] Comparing those inventories to Edward's demonstrates that many crucial implements were absent. A wheel-cutting engine, and metal-turning lathe with specialized accessories, would have been sufficiently valuable and distinctive to be itemized, as it was in David Rittenhouse's 1796 estate inventory where it was one of the highest-value items at £22.00. Edward's "large turning wheel" could have been a lathe, but not one suitable for clockmaking given its low valuation of just $2. If truly large it would have been for turning wood, not metal. Also absent were bellows, forms, and molds for casting brass plates, wheels, and other movement components.

Based on this it seems a reasonable assumption that Edward, upon leaving the city prior to the Revolution, also left his clockmaking behind, passing many tools and machines to his successor, John Lind, and choosing instead to spend his time managing his estate and pursuing other duties. Concluding instead that he never owned such equipment and never actually made clocks, simply assembling or retailing them, is a less-likely assumption.

*Detail of Catalogue No. 70 Surveyor's Compass. Photo by Gavin Ashworth.*

CHAPTER THREE

# Edward Duffield Clock Dials and Movements

*The Watch-Maker...scarce makes anything belonging to a Watch; he only employs the different Tradesmen amongst whom the Art is divided, and puts the several Pieces of the Movement together, and adjusts and finishes it... In treating of the Watch-Maker, I have said every thing that can be said of the Clock-Maker, or any other Branch of Tradesman concerned in making any Instruments for the Mensuration of Time.*

— Robert Campbell, *The London Tradesman: Being a Compendium View of All the Trades, Professions, Arts, Both Liberal and Mechanic,* 1747.

*The early American craftsmen were not inventors or improvers of clock mechanisms. On the contrary, they adhered strictly to the details of construction and styles of ornamentation current in the mother country.*

— Penrose R. Hoopes, *Connecticut Clockmakers of the Eighteenth Century,* 1930.

*Dial of Edward Duffield's standing timepiece (Catalogue No. 4), Collection of the American Philosophical Society.*

A typical eighteenth-century floor-standing clock, of the sort Edward Duffield made, was a timekeeping and hour-striking machine attached to an engraved and decorated dial displaying the hours, minutes, seconds, day of the month, and often phases of the moon. More complicated machines indicating tides also existed but were much rarer. Examples of these were produced by Franklin's English clockmaker, John Whitehurst, based on designs by James Ferguson. Philadelphia clockmakers Joseph Wills and William Stretch also made tide-indicating clocks, but none by Edward are known.

The true product of the clockmaker was the brass-and-steel machine hidden behind the dial. If the case bonnet had small, glazed side windows, the movement inside the dark cavity could be dimly observed but not closely studied without sliding off the bonnet. The clock's heavy brass dial was a multi-piece assemblage that in many instances was imported from England. Difficult skills involved in the making of these dials, separate from clockmaking, were mastered by specialists who could produce large smooth sheets, flat silvered rings, disks of painted or engraved stars and moons, subtle hammered patterns, and ornate cast-brass spandrels. Every clock made by Edward Duffield is an intricate and complex combination of materials and craftsmanship, but, while Edward may have done some of his own engraving, it is unlikely that he made the complex dials he signed.

By the time that Edward entered the trade, movements of eight-day long-case clocks had been standardized for nearly a century. There were small variations in wheel layout, part sizes and shapes, brass thickness, and steel quality, but nearly all movements had the same basic appearance, the same number of gears with the same or similar tooth counts, the same rack-and-snail striking system (only one countwheel-striking Duffield clock is known), the same pair of heavy iron or lead weights hanging below to provide the power, and the same kind of recoil or deadbeat escapement to keep the three-foot pendulum swinging exactly one second in each direction between ticks. A pair of rectangular flat brass plates, spaced apart by four or five brass pillars, enclosed the revolving wheels, rocking pallets, and hammer that struck the hours on a cast bell mounted above.

Unless a clockmaker was seeking ultra-accuracy or some unusual functionality, he had no reason to deviate from this successful design, and Edward was no exception. Clockmakers were craftsmen, not scientists, unless like David Rittenhouse they occasionally were doing more than building reliable domestic timekeepers. As Penrose R. Hoopes said, quoted above, they had no desire to deviate from the tried and true.

With four exceptions, Edward's movements adhere firmly to the standard format described above. One outlier was a thirty-hour one-hand version (No.53) recently examined in a private collection. This could be one of Edward's earliest efforts. Another outlier is the higher-accuracy, single-train timepiece (No.3) that Edward built for the APS's 1769 observation of the transit of Mercury. The third (No.4) is also at APS and descended through the Franklin/Bache families. Its movement is eight-day time-only and is one of four with a rare spherical-moon. Its smaller case and absence of distracting hour-striking made it appropriate for Franklin's library. The fourth (No.61) is another eight-day, single-train, smaller machine, without hour striking, housed in an unusual dwarf bombé case.

When movements from Edward's other clocks were available for even restricted examination, a few were noted that had "T" extensions from the tops of their front plates to accommodate the attachment of brass dials with large moon disks. One may be seen in Figure 3.1. Without those "wings" the toothed painted disk would have interfered with components mounted on the front plate. This modification was not unique to Edward, and sometimes just smaller "ears" extended out to attach the two top dial posts.

In general, when examining movements in Edward's clocks, we would expect them to be the same if he were making them from his own patterns and templates. As Robert C. Cheney explained, "… the movements made by Daniel Burnap (1759–1838) are readily recognizable. They agree in the planting of the train, design of the escapements, shape of the rack and rack hook and strike-lifting piece, and they agree, of course, with the surviving templates."[76] Instead, from movement to movement, Edward's are noticeably different and could point to different makers (Figure 3.1). For comparison, photographs of the front plates of movements from four other American and English makers from the same period are offered (Figure 3.2).

Close study of his movement plates shows variations in layouts of pivot-holes and front-plate posts that all must be precisely located. No clockmaker would want to repeatedly vary and compute their locations each time a movement was being built, especially after creating one that worked. Even a tiny error in locating and drilling pivot holes can lead to gears weakly meshing or jamming if too tightly engaged. A "depthing tool" is required at least for the prototype, and faint circles on some movement plates reveal that one was used.

Parts of the striking system, mentioned above by Cheney, are equally informative. The shapes and patterns of levers, racks, and lifting pieces on Edward's movements often are different from one another, rather than identical, as would be expected if he made them all from a proven pattern.

**Figure 3.1** Front plates of four Duffield movements showing the varying shapes of those steel parts and varying locations of front-plate apertures and pivot holes. Clockwise from top left: Catalogue Nos. 24, 32, 8, and 10.

However, some are the same and appear heavier and less decorative than parts typically seen on English-made movements, so perhaps he made these or had them crafted locally.

Some rack-tails on his movements have a springy steel finger that permits the clock to continue running if the strike train has stalled, or prevent damage if the hands are set backwards. This could be considered a possible marker for a Duffield movement, but four of nine examined movements do not have this feature, and other makers' movements have it as well.

The strike-system parts, with complex shapes and functions, were made mainly from steel that is far more difficult than softer brass for a clockmaker to forge, shape, cut, file, and polish. It is these parts which were more likely to have been bought in from different available sources. They must be precise to function flawlessly. If cut or placed incorrectly, they will not work. Trial and error, or remaking these parts multiple times until correct, was not a reasonable option for a busy clockmaker. Today's clockmakers can attest that fabricating such parts if missing, or correcting such parts that were improperly repaired or adjusted, is a tedious and frustrating task.

**Figure 3.2** Front plates of four eighteenth-century movements by two English and two American makers; (clockwise from top left) Joseph Harding, Samuel Mulliken, David Rittenhouse, and unsigned English (note its depthing-tool circles for accurate wheel placement).

1. Front Plate
2. Back Plate with Pillars
3. Weight Pulleys

**TIME TRAIN:**

4. Mainwheel with Click
5. Barrel with Ratchet
6. Barrel Clip
7. Center Wheel
8. Third Wheel
9. Escape Wheel
10. Pallets & Crutch

**STRIKE TRAIN:**

11. Mainwheel with Click
12. Barrel with Ratchet
13. Barrel Clip
14. Second/Pin Wheel
15. Third Wheel
16. Fourth/Warning Wheel
17. Fly
18. Hammer & Arbor
19. Hammer Spring

**ON BACK PLATE:**

20. Pallet Bridge
21. Bell Stand

**ON FRONT PLATE:**

22. Snail on Hour Wheel
23. Minute Wheel
24. Reverse Minute Wheel
25. Bridge with Hour Pipe
26. Setting Spring Clutch
27. Lifting/Warning Pieces
28. Rack & Rack Tail
29. Gathering Pallet
30. Rack Hook
31. Calendar Wheel
32. Calendar Wheel Post

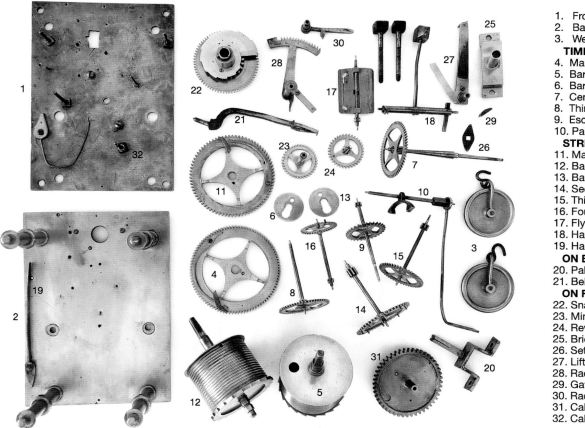

**Figure 3.3 (left)** Disassembled movement of Duffield clock Catalogue No.39. Author photo.

**Figure 3.4 (right)** Author's list of names of movement parts shown in Figure 3.3.

Also made from steel were the wheel shafts, or arbors, with cut pinions driven by the teeth of a larger adjacent brass gear. Forging, slitting, shaping, tempering, and polishing the pinion leaves was difficult and required specialized skills and equipment. Edward could locally purchase arbors with pinions, saving himself much time and labor and avoiding the necessity to own and maintain the required special tooling.

Only some movements of clocks in the Catalogue were available to clearly view, and only two were disassembled. Photos of one movement's multiple parts are provided (Figure 3.3) with numbering and naming of each part (Figure 3.4). The number of parts actually is greater than thirty-two since many are made from smaller components that normally are not separated. Also illustrated is the disassembled circa-1760 movement by Joseph Harding, a London maker of the same period (Figure 3.5) and side-by-side views of similar Harding and Duffield parts (Figure 3.6). Edward's gear rim and crossings are a bit beefier than Harding's but otherwise the wheels are nearly identical. "Beefier" is in any case not necessarily better; clockmakers would strive to make wheels light in order to reduce friction and stop-start inertia.

The hands and dials were removed from some clocks, when possible, to view the distinctive details on the front plates. More clocks visited in person could only have their case bonnets removed for side-views of the movements as shown (Figures 3.7 and 3.8).

Such photos typically are seen in auction catalogues and clock books and are minimally useful. They simply verify that a proper movement is present and is of the standard design. Such photos are not included in the Catalogue. An untrained viewer would discern no significant differences among the movements, nor would a professional in most instances. Edward's movements basically are like most others of his time, American and English. Claims are unconvincing that his movements are of higher quality or that there are appreciable differences in gear-tooth depth, brass thickness, finish quality, etc., between movements by Edward and those of his colleagues.

**Figure 3.5** Disassembled movement by Joseph Harding, London, circa 1760. Author photo.

**Figure 3.6** Comparisons of the same five parts from Duffield and Harding movements; Duffield's are on the left in each image. Author photos.

40

**Figure 3.7** (left). Duffield movements strike-side views, clockwise from top left, of Catalogue Nos. 12, 17, 26 and 27.
**Figure 3.8** (right). Duffield movements time-side views, clockwise from top left, of Catalogue Nos. 26, 6, 20, and 17. Author and commissioned photos.

The author has serviced hundreds of antique movements and sees nothing that would place Edward's work in a different or higher category. The movements and dials, whether made by him or outsourced, were comparable to most others of good quality. Edward needed to be competitive, but not innovative, for his upscale Philadelphia clientele, who could choose from many clockmakers, local and overseas. As noted by Jack L. Lindsey, "While many patrons sought to acquire 'the latest fashion' to demonstrate their taste, artists who wandered too far, too quickly, from the accepted models risked a loss of business or even ridicule."[77]

Edward's two disassembled movements revealed a shortcut that appeared on others by him as well. The flat outer surface of the front plate, rarely seen by the owner, was rough and had small fissures. Edward did not laboriously smooth and polish them, as was done on the other three plate surfaces. Most similar movements have all four flat plate surfaces bright and smooth, even if seen only by repairers. Very often those rough plate surfaces have a pink or orange hue, possibly a sign that copper in the brass alloy is concentrating on the surface. This color is not seen on polished brass surfaces, but sometimes also is noted on rough sides of brass weight pulleys. While fronts of English-made movements usually were smoothly finished, some were not, so rough coppery plates would not necessarily signal American origins.

Edward's dials are essentially the same as those signed by local and British makers. Figure 3.9 shows a London dial along with examples by Edward and his Philadelphia colleagues John Wood, Jr. and Thomas Stretch. Similar comparisons can be made with countless other dials of the period.

As just discussed, there are dissimilarities between Edward's movements—and, likewise, his dials are not exactly alike. This can indicate different suppliers over time. If he himself had been making his dials, there would have been no good reason for him to vary the dial features and components once he had created successful templates. Finally, as mentioned in the Introduction, his affluent and status-conscious customers often preferred buying imported English clocks, so to compete he may have assured them that the dials affixed to his movements were English, not home-grown copies that they may have considered inferior.

Mid-eighteenth-century advertisements confirm that these dials were being imported in quantity. As early as 1738, clockmaker John Wood, Sr. was recorded as purchasing five brass dials,[78] thus indicating that he was not fabricating them himself. In George B. Eckhardt's article, "Edward Duffield, Benjamin Franklin's Clockmaker," published in the March 1960 issue of *Antiques*, the author commented on the dial of an early Duffield clock (Catalogue No.53), "Like the dials of most clocks of the period, it was probably imported from England."[79]

At the same time, however, we do know from his dies for two commemorative medals that Edward had significant engraving skills, which he no doubt employed in his clock-making. He may sometimes have incised dial numbers and lines. Quite likely he engraved his own script signature, most often abbreviating his first name to Edw. or E and adding his city name in full or abbreviated form.

Engraved signatures and place names have been studied on fifty-two of Edward's clock dials and instruments. A sampling of his dozen round metal dial bosses, all of which were centered in the dial arch, is shown (Figure 3.10). In 32 signatures, he abbreviated his first name to "Edw." For nine he used his first initial "E." Four had "Edward" spelled out in full. On five surveying compasses, he omitted his first name entirely and engraved only "Duffield" at the top; on the other two he included "Edward" in a circle at the center. On thirty-five dials, he spelled Philadelphia's entire name. Seventeen others had Philadelphia abbreviated as "Philad.a." None of the abbreviations seemed required by available space; similar-size dial areas had the names fully spelled out, so we do not know why he abbreviated when he did.

In just four instances, for unknown reasons, he inscribed his information in block letters that look basically the same in each. All the rest were in ornate swirling script that also appeared done by the same hand and which matched his inked signatures.

If Edward, in fact, made the seven surveying compasses shown in the Catalogue, (Nos.64–70) they would demonstrate his highly skilled engraving. On the decorated dials, the finely engraved lines, numbers, and letters surround boldly drawn, crosshatched, and shaded foliate designs for the compass roses. These were traditional forms nearly identical to ones on English instruments from the same period and often copied by American engravers. For example, Benjamin Rittenhouse motifs duplicated many details found on instruments by Benjamin Cole Jr. (1695–1766) of London.[80] The complex engraving on Edward's compass dial of No.65 is mostly identical to one circa 1760, signed "Cole Fecit".

The ornate engravings on Edward's compasses are dramatically different on each instrument, however, perhaps suggesting that others produced them. Normally it was silversmiths and professional engravers who were the specialized artisans capable of the painstaking tasks of minutely and accurately dividing and inscribing the graduated scales with hundreds of precisely placed lines and numbers. This issue is discussed at length in the Catalogue introduction.

Eight special hand tools were needed for accomplishing the "common graduation" method of inscribing compass dials in Colonial America: "the dividing plate, an index, dividing knife, beam compass, spring dividers, a dividing gauge, a pattern scale, and a dividing square."[81] Edward would

**Figure 3.9** Comparable composite brass dials by Edward and his contemporaries, including Thomas Hunter of London, John Wood, Sr., and Peter Stretch. These further demonstrate the dials' standard and generic construction and appearance.

**Figure 3.10** Twelve dial bosses with Edward's engraved signature and city.

have needed to own these tools and to know how to competently employ them. They are not present in his estate inventory, something which also suggests that he did not do his own complex engraving. They may, however, have been divested decades earlier along with clockmaking tools and equipment also absent from the list.

As well, Edward never described or advertised himself as an engraver, and there were local professionals for Edward to call upon. Just below Edward's first newspaper ad in 1751, advertising his stock of olives and anchovies, was an ad by engraver Lawrence Hebert.[82] A 1760 ad related to the late engraver James Turner, located on Edward's Arch Street, included "Two Silver Watch Faces."[83] In 1764, Dunlap Adams advertised his many engraving proficiencies.[84] And in Lawrence Ash's 1762 advertisement in which he announced his repair services "Late from Mr. Edward Duffield's," he added that "Likewise engraving performed in all its Branches, in the neatest Manner, by John Steeper."[85]

Edward did possibly fabricate and fully engrave the less-complicated, silvered, one-piece, sheet-brass dials seen on just three of the Catalogue's clocks (Nos.3,15 & 38). Not commonly imported from England, these dials were large, thin, flat squares, two with added top arches. However, the arduous casting, hammering, flattening, scraping, and smoothing still would have been challenging muscular tasks that perhaps were done by other metalworkers, and unfinished imported sheet brass was available for purchase.

The sheet dials were engraved with numbers, lines, and patterns—far fewer than on compass dials—and filled with black wax to make them more legible. Sheet dials have been associated with clocks made later in the eighteenth century to compete with cheaper and brighter painted-iron dials being imported from Birmingham and affixed to locally made movements, but such less-expensive sheet dials already were in use in earlier years. The dial of the circa-1760 London movement by Joseph Harding, shown in Figure 3.5, is a modified sheet dial, a circa 1770 Simon Willard clock with a brass sheet dial stands at the Willard House and Clock Museum,[86] and Edward's clock No.3 with a sheet dial is firmly dated to 1769.

The dials of four clocks (Nos.4,10,17,19 in the Catalogue) have rare rotating spherical lunar indicators, with a numbered disk below, advanced daily by the movement. One vertical half of the metal ball was darkly colored, the other was brightly silvered (Figure 3.11). Although driven differently and without the horizontal numbered disk below the globe, some uncommon English clocks also have globular moons. Those dials were signed by Halifax makers Thomas Ogden and John Dixon, London makers Chandos Delander, Samuel Benson, William Dutton, John Ellicott, and Josephus Williamson, Manchester's George Booth, Ormskirk's Jonathan Taylor, and John Stephens of Bristol.

This author has not seen spherical moons on any other American-signed clocks, and they appear to have been a special, extra-cost feature that Edward could offer his affluent customers. Edward's signature styles and placements are different on each, and the applied standard corner spandrels are not identical. In each dial arch are brass castings with the dolphin motif. The original silvering is absent on the chapter rings, seconds tracks, and moon balls of two of these worn dials; the other two dials have had the silvering and blackened numbers properly restored.

Most of the English versions of these clocks rotated their globes via a vertical shaft rising from the movement front plate. Edward's simpler design used a rocking arm attached to the back of the dial, similar to the system for advancing toothed lunar disks found in most moon dials (Figure 3.12).

Perhaps Edward was inspired by an English clock, with a spherical moon on an angled shaft, which appeared recently at auction.[87] This clock is hugely important. It descended in Ben Franklin's family, and appears to be one of the clocks, with rare indicators of local tides, that Franklin commissioned from John Whitehurst of Derby, by whom the dial is signed, and brought back to America.[88] Whitehurst was a close friend of Franklin who visited the clockmaker's home in Derby more than once. Whitehurst also crafted the prototypes of the simplified three-wheel clock that Franklin invented and that was improved and described by James Ferguson (1710–1776).

**Figure 3.11** Four dials with the rare spherical-moon indicators. Note differences including signatures and spandrels. Clockwise from top left: Catalogue Nos. 17, 19, 10, and 4.

It was Whitehurst, not Edward, who made complicated and innovative clocks for Franklin. These included prototypes of Franklin's three-wheel movement as well as clocks showing local tides. More about this is in Chapter Nine.

Even more recently, a circa–1764 English bracket clock, signed by Samuel Northcote of Plymouth, came to auction.[89] Not only does it have the same small spherical moon in the dial arch, angled like Whitehurst's, but it has the same pair of tide-indicating subsidiary dials in the lower corners of the main dial. In this case, the movement retains the additional wheels, mounted to the movement front plate, for driving those two small dials and rotating moon. They are absent on the Whitehurst movement.

Multicolor painted-iron or enameled dials superseded both styles of engraved dials after the Revolution. Mass-produced in Birmingham, England, they were less costly. Their white backgrounds allowed easier viewing of the hands and numbers in typical low-light conditions.

**Figure 3.12** Backs of spherical-moon dials showing Edward's simple system for rotating the globes. Clockwise from top left: Catalogue Nos. 4, 10, 19, and 17.

There are no credible painted-dial Duffield clocks as might be expected if he had continued making clocks in the final quarter of the eighteenth century.

Were there poorer-quality movements than Edward's? Definitely, and especially if made in rural areas by craftsmen without access to better training, materials, components, tools, and affluent customers. Many such movements failed long ago, were discarded, and are no longer available for study. Low-grade clocks often are rarer than better ones, since they were not valued or kept by owners when the ticking stopped. Were there better and more inventive clocks than Edward's? Yes, and those came from England's finest makers and from Colonial artisans such as David Rittenhouse, who crafted more than standard domestic timekeepers. A magnificent example is the Rittenhouse astronomical musical clock at Drexel University. In contrast, Edward's single high-precision timepiece at the APS was well-made but unremarkable, except perhaps for its rate-adjuster. Nonetheless, his clocks are important and worthwhile to examine and to demonstrate his skill as a clockmaker.

CHAPTER FOUR

# Edward Duffield and Early Philadelphia Horology

*The working brasiers, cutlers, and pewterers, as well as hatters, who have happened to go over from time to time and settle in the colonies, gradually drop the working part of their business, and import their respective goods from England, whence they can have them cheaper and better than they can make them. They continue their shops indeed, in the same way of dealing, but become sellers of brasiery, cutlery, pewter, hats, &c. brought from England, instead of being makers of those goods.*

— Benjamin Franklin, *The Interest of Great Britain With Regard to her Colonies*, London 1760

*A treatise published in 1747 reveals that the English clock trade was organized in much the same way as the watch trade. Numerous specialized subcontractors performed repetitive tasks with precision and dispatch. The Liverpool industry relied on a vast network of "outworkers" to provide the components and services necessary to support thousands of journeymen clockmakers on both sides of the Atlantic. It was a factory-like system under many roofs rather than one.*

— Robert C. Cheney, *The Magazine ANTIQUES*, April, 2000

The Philadelphia horology story starts shortly before the end of the seventeenth century, when English Board of Customs records for 1697 indicated that £10 of wrought clockworks were exported to North America. The next year it was £20, and £115 in 1699. All of these were complete movements since export of parts was opposed by the London guild, the Worshipful Company of Clockmakers, that discouraged colonial clockmaking. In fact, a 1698 Act of Parliament "for the Exporting Watches, Sword-hilts, and other Manufactures of Silver,"[90] allowed complete watches to be exported if the maker's name was engraved on them, but prohibited the export of

*Dial of tall clock by Peter Stretch, private collection.*

empty watch cases which could be fit with inferior movements made in some other European country. Swiss and French watches were already illegally coming to America, but far fewer than those already being massproduced in Britain.

The earliest record we have of a tall-case clock being brought to Pennsylvania is one imported by William Penn in 1699. The clock was made by William Martin,[91] known later to have worked in Bristol, England, from 1703 to 1739, and stood in Penn's rural Pennsbury estate twenty-four miles up the Delaware River from Philadelphia. In the January/April, 2023, issue of *The Pennsylvania Magazine of History and Biography*, Maryland Institute College of Art professor Kerr Houston wrote in great detail about Penn specifically comparing governments to clocks. The colony was founded at a time of a leap forward in horological science and accuracy, with the application of the pendulum to clocks. Penn clearly was aware of these technological advances, and he noted that clocks, like governments, depend on good men making and running them. His 1682 fourteen-page pamphlet stated his intentions for the new colony's laws and contained relevant references to contemporary horological treatises.

In his authoritative 1884 history of Philadelphia, J. Thomas Scharf wrote that by 1700 the city had elaborately furnished houses with "tall Dutch clocks,"[92] suggesting that other immigrants had brought clocks with them as well. It seems more likely that the were English-made clocks, since Dutch clocks from that time were uncommon in England and North America.

Quaker records of bridegroom vocations in ninety marriages prior to 1708 list no young clockmakers or watchmakers.[93] The first known clockmaker to settle in the colony was Abel Cottey (1655–1711), who arrived in 1682 on *The Welcome* with William Penn. A 1709 movement by Quaker clockmaker Benjamin Chandlee (1685–1745) is Philadelphia's earliest known dated clock. Chandlee arrived in the city from Ireland in 1702 and apprenticed to Cottey, subsequently marrying his daughter Sarah. He lived in Nottingham, Pennsylvania, for the remainder of his life and was succeeded by equally skilled sons and grandsons.

Another early clockmaker, arriving in 1702 from England, Peter Stretch (1670–1746) in 1710 began maintaining a town bell actuated by mechanical clockwork in the Court House on Second Street. He was possibly its maker and likely oversaw its installation, and he was responsible for the clock until his death in 1746.

When Cottey died, Stretch was the logical choice to inventory his estate. He itemized parts, tools, and unfinished clock movements.[94] This itemized list demonstrated what a colonial clockmaker would need. Basic hand tools joined specialized ones made for casting brass wheels and plates, for laying out gear trains, and for drawing steel wire. In addition to the English parts already being imported, there may by then have been unfinished components available from local brass casters capable of providing them.

In 1717, clockmaker Anthony Ward (1669–1746) emigrated from Cornwall and was on that year's list of admitted city freemen.[95] This was the final year that new freemen were listed in the minutes of the Common Council. By 1724, Ward had moved to New York.

Despite the presence of local clockmakers, affluent citizens of the time often preferred London clocks and furniture to those made by local artisans. James Logan (1674–1751), William Penn's chief official representative in Philadelphia, ordered clocks in 1717 from London's Joseph Williamson.[96] Williamson (d.1725) was Watchmaker to the King of Spain, was associated with famous horologist Daniel Quare, was published by the Royal Society, and made a circa-1720 complicated clock movement[97] incorporating a spherical lunar indicator such as Edward made decades later. Quaker merchant Thomas Chalkley in 1720 imported two Japanned-case tall clocks from London, one blue with gilt highlights and the other colored black with gold.[98] Japanned furniture was regularly imported into the city at that time. Due to its surface fragility, though, few pieces in original condition remain today.

In the same year, in London, the Worshipful Company of Clockmakers petitioned for a ban on imported foreign watches. To bolster their case, the guild claimed that one hundred thousand clocks and watches were being made each year by British labor.[99] Many of these were sent to North American colonies.

Peter Stretch, in 1719, bought property on Second Street between Chestnut and Walnut, where he built his brick residence and shop just two blocks from Edward's future home at Second and Mulberry (later Arch) streets. Stretch lived and worked there until his death twenty-five years later. He possibly commissioned his first clock case in 1724 from cabinetmaker John Head (1688–1754), whose account book debited forty-one clock cases to Stretch's account, and showed a total production of approximately ninety-one clock cases. Four cases were charged to the account of Peter's clockmaker son, William.[100]

Stretch's home was typical for Philadelphia tradesmen. As described in a 1752 fire insurance survey,[101] it was seventeen-and-a-half feet by thirty-two feet, three stories high, with two rooms per floor. The first floor was divided into a twenty by twelve-and-a-half-foot-square kitchen and a snugger twelve-foot-square workshop in which he did his clockmaking.[102] Another fire-insurance survey from around the same period was conducted at the home of clockmaker William Huston. The property was valued at £225, divided between the house at £200 and separate kitchen at £25. The house, with no workshop noted, was: "10 feet front and 31 feet back - 3 storys high - 9 inch walls, plaster'd partitions, Board & Newel Stairs, outside and two storys within painted, garet plaster'd, a way out on roof and iron rails. The kitchen 10 feet by 9 feet, one story high, 9 inch walls. The whole quite new."

In 1720, a Market Street brass founder, John Hyatt, offered "brass work in the rough for clock-makers"[103] which would have freed city clockmakers from the hot task of casting gear blanks. The 1734 estate inventory of English clockmaker Edward Bullock indicated what was needed to cast such parts: "Flasks and Board ... Brass Heads (dial spandrel) Patterns ... Brass and Led & Wooden patterns ... Crucibles and Other Things."[104] Edward may later have sold such items, not listed in his estate inventory, or by the 1750s Edward's castings were readymade, bought either from England or from local brass founders such as John Hyatt.

Francis Richardson, Jr. (1705–1782) in 1738 sold his clock and goldsmith business to his brother Joseph and included two clock movements and five clock cases in the deal. Francis became a prosperous shipping merchant and active supporter of the Library Company and public hospital. The next year Joseph sold to Peter Stretch a "percell of clock movements,"[105] no doubt English imports. Francis later wrote from London that his brother should expect a shipment that included two moon-dial clocks and an additional clock dial.[106] Francis in 1744 again was in London placing orders that included "2 moon Clocks."[107]

Christopher Sauer (1693–1758), a Germantown clockmaker and printer, wrote in 1725 that he could build a clock in three or four weeks, and as this is similar to what other clockmakers reported, we have a sense of the time it would have taken Edward to construct one of his clocks. Christopher also admitted that he had no "cutting tool with which to cut gears"[108] so he was sourcing those parts locally or from England.

By the time Edward opened his shop in 1751, there were other clockmakers in Philadelphia who advertised their services in the *Pennsylvania Gazette*. These advertisements make interesting reading and from them it is possible to gain a detailed understanding of the broad variety of work performed by clockmakers in this period. Some advertisers announced their services cleaning and repairing clocks and watches, while others instead advertised their finished wares. In the latter case it is clear that much of what they sold was being imported wholesale from England.

In 1751, the same year as Edward opened his shop, Benjamin Bagnall, Jr., son of the famed Boston clockmaker, who had relocated to Philadelphia, advertised general wares including new silver watches and eight-day clocks with and without cases.[109] In the following year, Thomas Gibbons advertised: "Watchmaker from London, Next door to Mr. Inglis's, over the Draw-bridge, Cleans and repairs clocks and watches of all sorts, at reasonable rates."[110]

In 1754, watchmaker Henry Flower[111] advertised his shop very near to Edward: "At the sign of the Dial, in Second-street, Philadelphia, Choice Watch-Springs by the dozen, made in this city, and warranted to be good. The said FLOWER repairs, in the best manner, all sorts of Watches, whether of the plain, horizontal or repeating kinds; and likewise sets fine ENNAMELLED FACES in gentlemens or ladies watches that desire it."[112] In a postscript to a letter sent to Deborah five years later, Franklin reported that Flower's wife had appeared in London and appealed for financial assistance. Franklin noted that the woman was sent there ahead of her husband and that he was unsure about her identity and whether to advance more money. Nothing more is known about this sad situation.

"Horizontal" watches had newly invented escapements; plain watches had standard verge escapements. Costlier repeating watches sounded the time on small internal gongs or bells when a button or slide was pushed. White enameled dials were relatively new innovations that made viewing of the watch hands easier compared to older metal dials.

Also in 1754, Philadelphia merchant Benjamin Kendall at the corner of Third and Chestnut streets advertised goods imported on the *Beulah:* "to be sold cheap, for ready money or short credit… eight day and thirty hour clocks, silver watches, crown glass…"[113]

Two years later, Sansom & Swett, on Front Street between Market and Arch streets, announced imports via Captain Budden's *Philadelphia Packet* and other vessels.[114] Textiles and clothing were featured along with silver watches and eight-day clocks. William Ball on Front Street advertised in the same year: "A Parcel of good London made Clocks and Watches, the Maker's true name on them,"[115] while Henry and Woodham on Water Street listed silver and metal watches, indicating that lower-cost watches in base-metal cases were available.[116]

Merchant Joseph Richardson wrote in 1758 to his fellow Quaker Thomas Wagstaffe in London, ordering six silver watches to be obtained for £3 each from George Ritherdon (1753–1783).[117] Many Wagstaffe-signed clock movements were imported into Philadelphia and then installed in locally made cases. Joseph wrote in 1760 to Wagstaffe to report receipt of watches and enclosed an invoice for a: "Parcell of watches… 2 Silver watches at £5 or 5 Guinies Price Each…14 Silver watches at £0/55/0 a Piece…"[118] Again in 1762, he ordered from Wagstaffe: "a good Silver Watch of about 6 or 7 Guinies Price,"[119] and in 1766 he requested: "as many good five & six pound watches as the money will pay for."[120]

When clockmaker John Wood, Sr., died in 1760, his estate inventory included: "1 Prospect Time Piece, 14 Eight Day Clocks, 1 thirty hour Do unfinished, 1 allarm unfinished, 1 Clock case."[121] A "prospect" timepiece could have been a large clock easily viewed by the public, perhaps mounted on the outside of a building, or hanging in a shop window. The number of clocks showed that he was not building to order. If not imported movements, they would have represented many months of speculative labor without assured sales.

Advertisements placed in 1762 by Lawrence Ash are the sole indications that Edward employed an apprentice or journeyman. Ash announced: "CLOCK and WATCH-MAKER, (Late of Mr. Edward Duffield's) Having set up his Business in Front-street, six Doors above Market-street, would be greatly obliged to all Gentlemen and Ladies for their Custom and Recommendation; as it will be his chief Care to finish their Work with the utmost Dispatch."[122]

He advertised again in December of 1763, including an illustrated clock dial, but made no mention of having worked with Edward.[123] A few years later, a Lawrence Ash was working in the shop of Boston clockmaker Gawen Brown, perhaps suggesting that Edward's former associate had not prospered in Philadelphia.

**Figure 4.1**  John Sprogell advertisement, *Pennsylvania Gazette*, November 11, 1768, p.3. Courtesy of Newspapers.com.

In 1763, merchant William Ball advertised newly imported English goods at his Front Street store. Heading the long list were silver watches and eight-day clocks. Further down in the fine print were: "watch mainsprings, pendants, verges, hour and minute hands, watch glasses, clock dials cast or finished, heads and dolphins,[124] clock bells, and gut… a very large variety of best and common watch seals and keys."[125] This advertisement is especially significant as another indication that finished clock dials were available.

Another competitor within sight of Edward's door advertised his recent arrival in the city:

> John Ent, Clock and Watch-maker, Having set up his Business in Second-street, Philadelphia, between Arch and Race-streets … Makes and repairs all Sorts of Clocks and Watches, in the best and neatest Manner, as well as performed in London … He likewise undertakes Clocks by the Year, at a moderate price.[126]

In 1764, goldsmith and jeweler Edmund Milne advertised many new imported goods just arrived on the *Philadelphia Packet*: "At the Crown and Three Pearls, next door to the corner of Market-street, in Second-street… clocks in painted and siniered [veneered?] cases, neat gold and silver watches in engine turned cases, ditto in black and green shagreen cases studded… gold and silver silk watch firings, plain ditto, very neat gilt and steel watch keys and chains,.. all articles in the watchmakers business…"[127]

Early in life, artist Charles Willson Peale (1741–1827) taught himself clock and watch repair. He briefly advertised in Annapolis while working as a saddle-maker, claiming that he: "Makes, Cleans and Repairs CLOCKS, and Cleans and Mends WATCHES, in the best, neatest and cheapest Manner.…"[128] William Knapp, a more established Annapolis watchmaker, warned that watches serviced by others such as Peale: "have severely suffered thro' the unskilful and injudicious Practice of some Men, who assumed the Knowledge as a Business to which they were only Pretenders."[129]

Quaker Owen Biddle placed a newspaper notice when he moved from one house to another in the same year. The notice mentions repairing and making watches and also selling Trenton blistered steel used for purposes other than clockmaking: "OWEN BIDDLE, Clock and Watch Maker, is removed from the House next to the Corner of Market Street, in Third-street, to the House where George Owen lately dwelt, in Market-street; where he makes and repairs, All Kinds of Clocks and Watches, in the neatest and best Manner. Also Trenton blistered Steel, warranted to be of the best Quality, is sold in any Quantity…"[130]

In 1765 Samuel May and Richard Clarke advertised as "Clock and Watch-makers from London, at the Sign of the Hand and Watch in Front-street."[131] Joseph Bruff advertised two years later: "Goldsmith, at Talbot Court-House, Having procured a Hand from Glasgow, that has been regularly bred a Watch-Finisher, takes this Method to inform the Public, that he undertakes to make and repair either Repeating or plain Clocks and Watches."[132]

In 1768 John Sprogell at his newly opened shop on Front Street advertised with an eye-catching image of a clock dial (Figure 4.1), although most newspapers at that time had few or no graphic illustrations.[133]

Burrows Dowdney's ad of the same year, however, also drew attention with a picture of an arched moon dial:

> Watch and Clock-maker, in Front-Street, a few Doors above the Drawbridge, in the
> Shop lately occupied by Mr. Emanuel Rouse, Makes and repairs all Kinds of Clocks and

Watches, after the neatest and best manner.[134]

A mahogany bracket clock signed by Dowdney is in the Winterthur Museum collection.[135]

Freeman's auction house in Philadelphia sold (to the author) a 1772 legal document (see page 66), signed by Edward and several other city luminaries, addressing a land dispute involving the late Emanuel Rouse.[136] High Sheriff Judah Foulke issued this writ to seize and sell half of a house on Front Street formerly occupied by Edward's fellow clockmaker. Nathaniel Reynolds contested the sale, asserting that he held the title to the entire tenement and its ground.[137]

John Wood advertised again in the same year, but without a picture: "Clock and Watch-Maker, at the corner of Front and Chestnut Streets, has for sale, A Parcel of best Silver Watches; a variety of clock and watch tools; also watch main springs, fusee chains, glasses, enamel'd dial plates, pendants, pinion and pendulum wire, steel and brass keys, silk strings, steel chains and seals, small square steel, cast clock-work, ditto faces, finished and unfinished, heads, bells, clock and watch hands, case-string, catgut, &c."[138]

Mention of "cast clock-work" in this ad reconfirms that clockmakers bought some parts locally rather than making them themselves. In another advertisement Wood announced the services of newly-arrived English watchmakers:

> PHILADELPHIA MADE WATCHES. The subscriber having engaged in his employ some capital workmen from London, in the different branches of watchmaking, can furnish any gentleman with repeating, horizontal, seconds, or plain watches, warranted good… The advantage of having the maker of such machines on the spot is obvious.[139]

However, he still advertised, "Imported … Cast and forged clock-work, sheet brass, finished faces, cast watch-work, clock pinions cut…"[140]

Further evidence that more clockmakers were settling in the colonies is provided by a notice that a ship, *Jupiter*, arrived in 1773 from Londonderry with "A FEW IRISH SERVANTS, whole indentures are to be disposed of; among which are two good clockmakers…"[141] A 1771 letter from Owen Biddle, who wrote to England indicating that he was importing clocks but would cease doing so since business conditions were poor and competitors were flooding in, also provides evidence of this increasing competition:

> Notwithstanding the Clocks thou sent were exceeding Good in Quality, yet such is the small demand for them as well as Watches at this time, that I have no encouragement to order any now, many Workmen having come from Great Britain in that Branch of Business, who execute their work well and afford it as low as they can be imported. This is like to be the case, with most branches of Manufacture, which we are not supplied with from your parts in Course of time as we have every advantage for such an undertaking, except hands, which increase amazingly daily.[142]

Biddle also wrote to Thomas Wagstaffe complaining of English watches' low quality, and saying that he could not recommend them.[143] Perhaps Biddle received poorly made watches that could not be easily marketed in the home country. Overseas customers commonly asserted that products made for export were inferior since complaints and returns were difficult or impossible.

Other advertisements from this period include one by goldsmith John Dupuy, who was located "in Second-street, below the Friend's Meeting-house," and advertised that he "made and mended, all kinds of clocks and watches,"[144] and a request in 1770 for owners of watches left at the shop of Thomas Stretch, who had died five years previously, to pay related charges to Isaac Stretch, "Watchmaker in Second-street" or the watches would be sold. Maker names on those watches were listed: "David Terrier; Jaques Poette; Benjamin Bell; Anthony Ward, Philadelphia; and Peter Semaistre."[145]

Samuel Jefferys advertised his work:

WATCHES. JUST IMPORTED from London, in the Britannia, Captain Falconer, A Parcel of new Watches, warranted to keep good time, and to be sold at very moderate prices. Clocks made and mended; also repeating, horizontal and plain Watches repaired in the compleatest manner, by Samuel Jefferys, in Second-street, between Christ-church and the Court-house, Philadelphia.[146]

Merchant Mary Eddy advertised: "A neat assortment of clock and watchmakers tools, watch faces, key, seals, and hands, inside and outside chains, a variety of watch glasses, pendants &c. clock and, gold lacker, clock movements, finished and unfinished, with and without faces, and cat gut for clocks, &c."[147]

In 1773, John Wood, Jr. advertised many imported goods including "SILVER WATCHES, of various prices; cast Clock-work, with or without faces, at 2s. 10d. per pound, corner and arch pieces, crucibles, clock wheels and pinions, turned and cut… barrel arbors and ratchets,… finished clock faces,… forged clock work…"[148] and many more tools and parts for working horologists. This demonstrates the wholesale availability of finished clock dials as well as clock components by the pound.

Merchant Timothy Berrett, near Edward, "In Arch-street, between Second and Third-streets," again in 1773 offered recent imports: "two very neat enamelled arch dials for clocks, a few sets of clock work, casting and forging…."[149] This is a significant early notice of painted dials arriving in Philadelphia. These soon superseded the costlier and harder-to-read brass dials found on Edward's clocks. Painted dials were first advertised in Birmingham, England, the previous year by Osborne & Wilson who pioneered their development.

Laurence Birnie set up shop within view of Edward's establishment, although the Irish immigrant was soon occupied as a captain in the local militia:

WATCH and CLOCK MAKER, from the city of Dublin, Begs leave to acquaint his friends, and the Public in general, that he served a regular apprenticeship to his father William Bernie, late of Temple-Patric, and afterwards experienced a considerable share of knowledge with some of the most eminent Watchmakers in England and Ireland; and as he has opened a shop at his lodgings in the house of Mrs. Faries, in Arch-street, near Second street, where he intends carrying on said business in all its various branches…[150]

The next year, Birnie established a file-cutting mill and furnace in connection with a new gunlock factory. He was one of several horologists, not including Edward, whose skills were diverted to Revolutionary War efforts.[151]

Clock and watch repairs continued during the long years of the Revolution. White and William Matlack, half-brothers of brewer and Philadelphia politician Timothy Matlack, advertised in 1780: "All kinds of CLOCKS & WATCHES MADE or REPAIRED… The South side of Market-street near Fourth Street."[152] A 1782 advertisement noted the clock-dial trade sign for customers visiting a shop quite near Edward's corner: "Thomas Morgan, Watch and Clock-Maker, At the Arch Dial, in Arch-Street, between Moravian-Alley and Third Street…[153] Morgan's manuscript ledger may be viewed at the Historical Society of Pennsylvania.[154]

William Reeves scratched a 1783 cleaning date on the movement of the Duffield clock at Colonial Williamsburg—Catalogue No.19—while 1798 was inscribed later by another clockmaker Samuel Norton. Edward at Benfield was not hired for the service of a clock he signed, again suggesting he had retired from the trade.

When peace between England and the new United States finally arrived in 1783, a letter from the Richardsons provided evidence that importing clocks from England had resumed. They ordered from Masterman & Son in London but were unhappy about the high prices: "2 good 8 day Clocks with plain inameled faces without Cases… The Clocks dont quite answer our expectations owing to our not being more particular in describing them, they came very high being twice as much as we had any idea of."[155]

Ephraim Clark, whose Philadelphia shop reportedly was later visited daily by George Washington to set his pocket watch, advertised "Neat Silver Watches… to be SOLD cheap… Watches repaired as usual."[156] In another advertisement, Clark offered newly arrived goods including: "EIGHT-DAY, moon, and plain arched clocks… Japanned clock faces…"[157] M.F. Tennant wrote that Clark, "was importing cases of Birmingham painted dials from Liverpool by 1790,"[158] indicating that painted dials were available in quantity.

John Wood, Jr. posted prices for varieties of clock dials in a 1787 trade catalogue:[159]

| | |
|---|---|
| 13 Inch Japanned Moon Faces | 2.1.2 |
| Ditto for Seconds in the Center | 2.1.2 |
| 12 Inch Jappan'd Moons | 1.15.1 |
| 12 Inch Solid Arch Japan'd Faces | 1.1.7 |
| 11 Inch 30 hour Ditto Arched thirty hour | .18.5 |
| Finished Brass Faces moons | 2.16 - |
| Solid arched Do. | 2. - |
| 12 Inch Arch Faces China Enamald with Flowers | (price indistinct) |
| 13 Inch Do. | 20/ to 24/ |
| 12 Inch Do. with moons do. | 24 to 32/ |

Historian Daniel Boorstin wrote about the diversified English watchmaking trade at this time:

> In England watchmaking was showing the advantages of specialization and the division of labor. The Clerkenwell district of London housed various groups of workmen who called themselves escapement makers, engine turners, fusee cutters, secret springers, or finishers. The Clockmakers' Company reported to the Board of Trade in 1786 that they were exporting about eighty thousand clocks and watches each year to Holland, Flanders, Germany, Sweden, Denmark, Norway, Russia, Spain, Portugal, Italy, Turkey, the East and West Indies, China, and elsewhere.[160]

Oddly, the United States in general, and Pennsylvania in particular, were not mentioned as receiving these exports although they clearly were. Bache (Franklin's son-in-law) & Shee were among those importers. In 1786 they posted a long two-column list of goods for sale including "Chamber clocks."[161]

Advertisements were placed not only for the sale and repair of clocks or watches. Some appeared for lost or stolen watches, and these are particularly informative about makers and styles. Richard Smith posted: "Lost in the Hurry of Removing Goods at the late Fire, a large striking Watch, with a brass inside Case and a silver out-side Case with round Holes to let the Sound through, both Parts of the movement going by one Spring, R. Arnold, Providence, engrav'd upon the Dial Plate…."[162]

Such striking—or repeating—watches were uncommon and costly. A woman attempted to pawn a watch case that, when taken to a silversmith to be valued, turned out to be Smith's. She confessed to the theft and was jailed.

Peter Stretch advertised for a lost or stolen watch by Strowd of London.[163] He added that it had a key and seal and black silk string, a small gold rose on its dial between every hour and half hour, and graduations for the minutes. Seals often dangled in plain sight from the same ribbon or chain as watch-winding keys. Another 1755 advertisement alerted readers: "Lost between Bush-Hill and Philadelphia, on the 18th of August, a small sized Watch, with a shagreen case, maker's name James Good, London, No. 572. Whoever finds said watch and brings it to Mr. Stretch, watchmaker in Second-street… shall have Twenty Shillings reward. If offered for sale, it is desired to be stopt."[164]

In 1769, a £3 reward was offered for a lost watch: "on the Middle-Ferry Road, a double cased Pinchbeck Watch, engraved flowered Case, opens in a very particular Manner, Maker's Name GRANTHAM, London No. (supposed to be) 2395, a Steel-studded Chain, Steel Watch Key, and… a small Locket, with Hair in it."[165] Human hair keepsakes were common. Pinchbeck was a cheap gold-color alloy of copper and zinc named for its inventor.

A London Wagstaffe–signed silver watch was lost in 1767 in Lancaster or on the road from there to Wright's Ferry. An ad placed in Philadelphia noted that the watch had "a short steel chain, with a pinchbeck seal, and two keys, the one brass, and the other steel… All watchmakers are requested to stop the said watch, if it should be brought to them."[166] "Stop" meant retain the piece, not halt its ticking.

Lost watches were advertised in 1771: "Was found in Bristol, by a Passenger in the New-York Stage, a remarkable SHAGREEN CASED WATCH."[167] A two-dollar reward was offered on the same page: "DROPPED, last Friday Afternoon, in Fourth-street, between Walnut and Market Streets, an Outside SHAGREEN WATCH CASE, with a Chrystal on the Back of it." Shagreen is sharkskin or leather with a rough outer finish. A rear glass crystal is an unusual feature known as an exhibition back which permits viewing the watch movement without exposing it to dust.

A generous four-guineas reward was offered in 1777 for an unusual lost silver watch signed: "Charles Geddes, New-York; she is of a singular construction, having no figures for the minutes on the dial-plate, which is the only one of the kind, with a cap, of Mr. Gedde's make in America…."[168]

Philadelphian Tench Coxe advertised in 1779 for a watch lost by John Coxe in Bucks County. Watchmakers were asked to be vigilant for: "A PINCHBECK WATCH, wrought and flowered on the case, which does not separate from, but is joined to the body of the watch by a hinge; the watch was made in London, maker's name Baird."[169] A sixty-dollar reward was offered, an indication of worth but also of war-time currency inflation.

Some clocks were stolen during the war by British troops and after the British retreated from Philadelphia. These pilfered goods, including two locally made clocks, began to surface in advertisements: "Was purchased last winter of the British soldiery, for a trifling consideration, by a young Gentleman of this city, with a view of restoring them to their proper owners, two eight-day CLOCKS, makers names Anth. Ward and Peter Stretch. The owners proving their property and paying a few inconsiderable charges, may have them again."[170]

Philip Ling "in Vine-street, near Second-street" had a looted Stretch clock, stating the exact date of the theft:

Was brought to this city on the 11th of December last, by a part of the British army, an eight-day CLOCK; it shews the day of the month, and when taken shewed the 11th of December, maker's name Peter Stretch, Philadelphia: there were brought with it two weight-wheels and one weight, the pendle and a piece of the wire with the bell crack'd: it was offered for sale at a low rate, and was purchased by an inhabitant of this city.[171]

Another good citizen provided a list of items including an eight-day clock by Owen Biddle, and a roasting jack that had been abandoned when British soldiers were approaching.

Philip Wagner, Captain of the Fifth Pennsylvania Battalion of the Philadelphia Militia, offered an eight-dollar reward for the return of a locally-made clock:

[T]aken out of the house of the subscriber, in the Northern Liberties, between Germantown and the Rising Sun, on the Philadelphia road, by some of the British troops, on the 25th or 26th of September, 1777, a repeating 30 hour clock, with an alarum, minute hand, and day of the month, the maker's name Augustine Neisser, engraved on the circle, the face eleven inches square: It was taken without the pendulum and weight.[172]

Alongside these advertisements, trade journals, catalogues and articles of the time give us insight into how clocks and watches were made. The 1754 Volume Twenty-Four of *The Gentleman's Magazine and Historical Chronicle* had several pages of text and illustrations describing timepiece escapements including innovations by French horologist Jean-Andre Lepaute (1720–1789). The escapement, the ticking heartbeat of a clock or watch, converts the force of a rotating toothed gear into the back-and-forth swinging of a pendulum or balance wheel. All Edward's clocks used "anchor" escapements. Some were "deadbeat," more difficult to make but considered more accurate.

**Figure 4.2** Clockmaker's center lathe, Plate 22 of John Wyke Catalogue of Tools for Watch and Clock Makers. Author photo.

The fifth edition of *The Artificial Clockmaker* by William Derham (1657–1735) was published in 1759. In 1696 this classic text was one of the first books to describe horological arts in detail. An ordained clergyman, Derham produced many other scientific treatises and was the first to accurately determine the speed of sound. Edward may have studied the new edition that added a fold-out equation-of-time table and coverage of the Gregorian calendar belatedly adopted by England and its colonies.

The third edition of Thomas Chippendale's *The Gentleman and Cabinetmaker's Directory* appeared in 1762 in London and arrived the following year in Philadelphia with immigrant cabinetmaker Thomas Affleck. Elegant designs included clock cases that influenced those made for Edward. As noted by British horological author Deryck Roberts, however, Chippendale's actual designs were impractical: "Chippendale's suggestions for clock cases demonstrated a total lack of knowledge about the workings of clocks... required specially-made dials... were too ornate... they would be a disaster."[173]

In the late eighteenth century, John Wyke (1720–1787) of Liverpool published an illustrated *Catalogue of Tools for Watch and Clock Makers*. He offered full selections of tools and equipment. Specialized implements were available to the trade well before this catalogue was printed. Nearly a century earlier, English horological inventor Robert Hooke wrote of a wheel-cutting engine, as did a London watchmaker twenty-five years later who described hand-powered engines that would cut wheel teeth, adjust balance wheels, cut grooves in fusees, and draw steel pinion wire. The Wyke catalogue was reproduced in 1978 by the Winterthur Museum Library, with extensive text by curator Charles F. Hummel. The full-page plates demonstrate that eighteenth-century horologists required and could purchase sophisticated equipment.

David Rittenhouse died in June of 1796.[174] Two articles of his were published posthumously in 1799 in Volume Four of the *Transactions of the American Philosophical Society*. The first, *On the Improvement of Time-keepers*, transcribed a 1794 lecture and began: "The invention and construction of time-keepers may be reckoned amongst the most successful exertions of human genius. Pendulum clocks especially, have been made to measure time with astonishing accuracy; and if there are still some cases of inequality in their motions, the united efforts of mechanism, philosophy and mathematics will probably in time remove them." The second was a 1795 letter that described his experiments measuring the effect of heat on wooden pendulum rods. Edward is likely to have read both of those treatises.

Other documents give us insight into which watch and clock parts a clockmaker such as Edward might have purchased locally. We know that Philadelphia silversmith Joseph Richardson (1752–1831) engraved dials for clockmaker John Wood, Sr., in 1734,[175] and that in 1737 he charged Wood for: "Caseing a watch Case silver & making"[176] suggesting that he had the required skills. Brass–

founder John Stow (1726–1754) in 1752 advertised a long list of products concluding with, "work in the rough, for clock-makers, &c."[177] James Smith and John Winter advertised their Philadelphia bell–casting business: "as good and as neat as any that come from England."[178] These men could have supplied bells to Edward and other local clockmakers.

Other advertisements mention clock glass which clockmakers certainly could not make themselves. A local merchant to, "Next Door to the Sign of the Bird in Hand, in Chestnut street, near Peter Stretches corner," advertised London crown glass suitable for clocks.[179] Glazier Thomas Ellis advertised: "London crown ditto, of any size, fit for clocks…."[180]

Such old crown glass, with its distinctive lines and imperfections, can still be found in the bonnet doors of antique clocks. Clock cases, cabinets, and wall brackets were of course ordered from local cabinet and furniture makers. John Wood, Sr., ordered a clock case in 1750 from cabinetmaker Henry Clifton, bartering with mahogany boards,[181] while Hornor's *Blue Book* described wall-mounted carved wood brackets made by city cabinetmakers to support small shelf clocks. Hornor recorded that Thomas Affleck made one for Joseph Pemberton at a cost of fifteen shillings. He added that these "were secured at the correct point of vision, and yet well out of the way of curious children."[182] Small clocks on wall brackets were appealing less-costly alternatives to floor-standing models.

While these documents provide us with evidence on clockmaking as a trade, others provide insight into how one became a clockmaker and what was expected of an apprentice in the trade. In 1741, Peter Stretch signed a five-year apprenticeship indenture with Emmanuel Rouse, son of widow Dorothy McNeal. Typical of those wordy and ornate documents, young Rouse's indenture mandated that he would faithfully serve and obey and keep his master's secrets, do no damage or allow others to do so, nor waste or lend his master's goods. Fornication, matrimony, cards, dice, ale-houses, taverns and play-houses all were forbidden. Six years later, a 1747 advertisement by silversmith Elias Boudinot (1706–1770) stated that at his address, "clocks are made after the best manner, and watches are repair'd, by Emanuel Rouse,"[183] so Rouse was apparently well trained. He went on to have his own shop as well, for twenty years later, clockmaker Burrows Dowdney advertised his location "in the shop lately occupied by Mr. Emanuel Rouse."[184]

David Rittenhouse, whom John Adams described as a "tall slender man, plain, soft, modest, no remarkable depth or thoughtfulness in his face, yet cool, attentive, and clear,"[185] mainly made his living as a clockmaker from around 1750 until the Revolution, the same years as Edward, and the two men certainly were well acquainted. Like Edward, he came from Philadelphia's affluent classes, as his great-grandfather, Willem Rittinghuysen, built America's first paper mill and added sawmills and gristmills. How Rittenhouse learned clockmaking is unconfirmed, but as with Edward, he never served as a lowly apprentice. J. Thomas Scharf claimed that "there was already a trunk full of tools appropriate for the trade in the garret, the property once of some maternal relative",[186] so perhaps his interest in clockmaking grew from this.

His clockmaker brother, Benjamin (1740–1825), however, did take apprentices. In 1781 he directed new applicants to David's home:

> WANTED, AN ingenious Lad not exceeding 14 years of age, of a reputable family, as an Apprentice to learn the Art and Mistery of making *Clocks* and *Surveying Instruments*…
> Any lad inclining to go an apprentice to the above trade, the terms on which he will be taken may be known by enquiring of Mr. David Rittenhouse, in Philadelphia, or at the subscriber's house in Worcester township, Montgomery County.[187]

Other clockmakers, perhaps to save money, combined an advertisement for their wares with a search for an apprentice. John Stillas announced the opening of his shop "in Front Street, second door above Chesnut-street." He added that "An ingenious Lad will be taken as apprentice."[188] Another man, advertised new merchandise from London at the same time as he sought a new apprentice: "John Riley, Clock and Watchmaker, Seven doors below Market-street, on the east side of Second Street, a neat ASSORTMENT of Capped and plain Silver Watches… Likewise second-hand Gold and Silver WATCHES. CLOCKS and WATCHES repaired, as usual… A smart active BOY, who can be well recommended, will be taken as an Apprentice."[189]

John and Daniel Carrell advertised both their merchandise and their need for an apprentice: "Gold and silver watches, eight-day clocks, repeating spring clocks…"[190] They sought apprentices and sold "cast clock-work in setts: clock-bells; clock pinions, ready slit…."[191] This is a noteworthy mention of sets of rough castings and forgings ready to be finished into complete movements.

Journeymen—successful apprentices who worked for wages from a master—were also sought. In 1769, an anonymous advertiser desired: "A Journeyman Clockmaker, of a good character, who understands his business, may hear of constant Employ, by applying to the Printer hereof."[192]

Advertisements also reveal that not all apprentices were able to complete their training with the same master. Benjamin Rittenhouse was ailing in 1779 and could no longer employ his apprentice: "TO BE DISPOSED OF, THREE years time of an handy apprentice lad, who has been bred to the gun-smith's and clock-maker business. His master's health will not admit of his following his occupation, which is the reason why he is disposed to part with him."[193]

Nor were all clockmaking servants and apprentices successful employees. Some went missing while others could not resist the temptation of stealing the valuable watches they were learning to make and repair. Clockmaker Isaac Pearson, for example, reported runaway servant-clockmaker John Williams who "speaks by Clusters, hard to understand…"[194] while Peter Stretch reported a missing apprentice named William Cannon: "of a middle Stature, short brown Hair, having on a blew gray Suit of Broad Cloth."[195] Wilmington, Delaware, clockmaker George Crow offered in 1752 a £5 reward for the return of his runaway English indentured clockmaker named Henry Bimpson. A full description was provided of his appearance and clothing that included an old weather-beaten wig.[196] Volarius Dukehart of Baltimore advertised in 1770 in Philadelphia for the capture of a bail-jumping watchmaker, Barsel Francies. A £15 reward was promised. He was "very remarkable in his Walk, both of his Ancles (sic) being very crooked."[197]

On occasion, an advertisement revealed that it was the master who got into trouble. In 1766, Thomas Skidmore from Lancaster advertised in Philadelphia for an apprentice:

Wanted, an Apprentice to a Watch-maker; he must be a Lad of genius, and of creditable Parents; he must serve Seven Years, and notwithstanding he will have an Opportunity (which is not very common in America) of making the Movement, and finishing the same, the Apprentice Fee (if small) provided the Boy suits, will be accepted. Any Person having such a Boy to put as an Apprentice, may send to Thomas Skidmore, Watch-maker, in Lancaster, who is the Person that wants the Boy. N.B. The said Skidmore learnt his Business in London.[198]

In the following year he advertised that his shop was making London-quality watches:
[For] £12 currency; and as many people have been under the necessity of importing good watches from England or Ireland, he now assure those who would incline to have good watches that he has, besides himself, two regular bred workmen from England; the one a movement-maker, and the other a motion-maker: Therefore any person who wants a watch, he will warrant it for three years, without ending, to the purchaser, and whatever size the purchases chooses, from the size of a half dollar, to a larger, they may have, and in three months time from bespeaking of it…. He also makes wheels, pinions, verges, &c. for watch-menders, which he will sell low. Any person wanting a watch of Twenty Guineas price, may be here supplied as in London.[199]

However, a few months later, Christian Voght alerted the public about the same Skidmore: Run away from his bail, on Friday Last, a certain man, who called himself THOMAS SKIDMORE; he is a watch-maker by trade, is about 5 feet 3 or 4 inches high, and well set; has brown hair, and carries a very bold countenance…; he took with him a grey gelding, belonging to Dr. Stuber in this town…. Whoever takes up the said Thomas Skidmore in this province, and brings him to his bail, shall have Ten Pounds reward; but if taken in another province, Fifteen Pounds reward…[200]

Pennsylvania artisans and workers rarely wrote their stories. An exception was William Moraley (1699–1763). With only clockmaker skills, he was the final immigrant waiting at the Philadelphia docks to be indentured. He was sold for just £11 to clockmaker Isaac Pearson (1680–1713) of Mount Holly, New Jersey. Moraley published his saga in London in 1743. He had apprenticed under his watchmaker-father, who trained with one of England's most famous horologists, Thomas Tompion (1639–1713). Moraley frankly revealed his dissolute life in England and America despite having been trained in a prestigious trade.

Moraley described his life while indentured to Pearson, who typically needed to augment his income by farming and working as a silversmith, goldsmith, blacksmith, and button-maker. Pearson dispatched Moraley around the countryside, and the worker closely observed the area's inhabitants, fauna, and flora. He recorded his master's stiff competition, naming Peter Stretch as the preeminent clockmaker along with John Wood, Sr., and Edmund Lewis. Wood and Stretch are well-known but not Lewis, who Moraley described as a young Quaker who loved supernaculum (good-quality alcoholic beverages). When his indenture was fulfilled, he worked briefly in Philadelphia for Lewis and for William Graham, nephew of famed London watchmaker George Graham, but he then led an itinerant life before returning to England and drafting his memoirs.

Clockmakers not only made clocks and watches used by people in their daily private lives. They also made clocks to be displayed publicly on buildings, created other forms of clockwork mechanisms and invented new methods of constructing clocks. Thomas Wagstaffe, in 1764 donated a spring-powered gallery clock to the Pennsylvania Hospital. Wagstaffe's clock arrived from London with the accompanying letter:

Esteemed Friends, — The Regard I bear the Province of Pennsylvania, Respect to the City of Philadelphia in Particular & Esteem for its Inhabitants. The Distinguishing marks of the Favours I have received from them Claim my acknowledgements and as a small Token thereof Present them with a Spring Dial for the use of the Pennsylvania Hospital to be fixed up therein at the Direction of the Managers. In the Performance whereof I have not so much Consulted Ornament & Elegance as real Usefulness being executed in the best Manner for Keeping Time. I request your acceptance thereof and am with Real esteem Your Assured Fr'd, Thos. Wagstaff. London the 16th 8th Mo. 1764.[201]

**Figure 4.3** 1785 movement formerly in a Newburyport meeting-house steeple. Author photo.

In 1753 Peter Stretch's son, Thomas, who was a competent clockmaker like his father, completed and installed the State House clock that later was maintained by Edward. Significantly, the clock's movement was constructed locally, not imported. In 1759, Thomas finally was paid £494.5.5 "for making the Statehouse clock, and for his Care in cleaning and repairing the same for six years."[202]

This author viewed a tower-clock movement, made by an unknown American or English craftsman, that most likely is similar to the machine Thomas constructed (Figure 4.3). Now in a private collection, it was purchased in 1785 for installation in the steeple of the meeting house of the First Presbyterian Society in Newburyport, Massachusetts. Its wrought-iron frame construction was standard for that period. Thirty-five inches across at the base, it probably was somewhat smaller, however, than what Thomas would have needed to drive the pair of large hands on the State House building's exterior. The Newburyport clock's care during more than a decade after installation was in the hands of well-known clockmaker David Wood (1766–1855) of that town, as documented by several dated receipts.

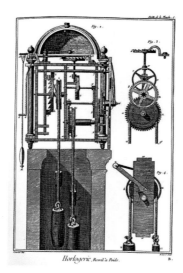

**Figure 4.4** Diderot drawings of a lantern clock with alarm. Author's collection.

Thomas Stretch also was a governor of the State in Schuylkill, the city's exclusive fishing (and drinking) club still in existence. In 1763 Edward was not listed among the 113 members, who included many of his city friends and associates: Henry Drinker, Joseph Fox, William Bradford, John and Thomas Potts, William Plumstead, and fellow clockmaker Jacob Godshalk.[203] After his father's death in 1746, Thomas Stretch sold the shop on Front Street and moved near Edward at the southwest corner of Second and Chestnut Streets.

Philadelphia watchmaker Robert Leslie, often at odds with other horologists in the city, invented new methods of suspending clock pendulums. The Assembly, as part of its efforts to encourage domestic manufactures when British imports were unavailable, awarded him a patent in 1777. Leslie cared for the State House clock during the late 1780s, and he died in 1803, the same year as Edward.

English clockmakers also made elaborate clockwork instruments that attracted special interest when toured overseas. The first of these in Philadelphia was seen at the house of Mr. Videll in 1744, when the public paid to enjoy performances by a large musical clock crafted by British clockmaker David Lockwood. The tunes' many composers included Frederick Handel. According to ads placed in the *Pennsylvania Gazette*,[204] the King in London had complimented the machine that was for sale along with a large camera obscura. Guests could examine the complex clock movement, and this would have intrigued teenage Edward who lived close by. Today, Lockwood's masterpiece is lost.

In that same year, the French-language comprehensive encyclopedia of world knowledge by Diderot and d'Alembert was launched. It later included more than 250 articles on horology along with seventy related engravings. One illustration showed a typical lantern clock that Edward did not make but could have owned and serviced (Figure 4.4). The "alarm" listed in Edward's estate inventory may have been one of these.

Another special large clockwork device, the "Microcosm: Or, The World in Miniature,"[205] was displayed in 1755 at the Front Street home of merchant Charles Stedman during its North American tour (Figure 4.5). Edward could hardly have resisted viewing this mechanical wonder built in 1734 by English clockmaker Henry Bridges of Waltham Abbey. It stood ten feet tall and boasted 1,200 wheels and pinions. Music played automatically on several instruments to accompany the animated performances of mechanical people, animals, birds, ships, and vehicles.

**Figure 4.5** Microcosm, *Pennsylvania Gazette*, January 29, 1756, p.4. Courtesy of Newspapers.com.

This machine, viewed by a young George Washington during its display in New York in February 1756, disappeared long ago but a small portion was rediscovered and is exhibited at the British Museum.[206] Paul E. Sampson explained how the Microcosm illustrated for average citizens the Newtonian concept of the universe:

> With the turn of a key, the entire cabinet jolts to life…. At the peaked top, automaton figures of the muses play instruments in concert on Mount Parnassus. Below, an artificial Orpheus performs for an audience of wild beasts that nod and sway to the music of his flute. On the bottom tier of the case, two more animated scenes display a prospect of the ocean in which ships sail while coaches and carts hurry along as the wheels turn round as if actually on the road. Finally, at the very bottom, you behold a very entertaining representation of a carpenter's yard in which animated laborers saw and hammer tirelessly. At the center of all of the frenetic activity is a clock. But an extraordinary clock: the dials display not only the time, but also the motions of the Ptolemaic and true Copernican universe, including the precise locations of the six planets.[207]

Shortly thereafter, in England, John Harrison completed his fourth marine timekeeper after five years of intense labor. Prototype and predecessor to later indispensable ocean-going marine chronometers, it was tested during a 1761 ocean voyage to Jamaica, and Franklin sat on the Council of the Royal Society that advised the Board of Longitude about equipment the ship should carry.

David Rittenhouse constructed an observatory for the June 1769 transit of Venus. The astronomical regulator clock he used during the event is in the collection of the APS (Figure 4.6).[208] He stated after the transit that he had not finished making his equal-altitude instrument and thus could not set his timepiece exactly to noon but he still could use it for timing observations. He also did not have time to fabricate a temperature-compensating device for its pendulum. He described crucial moments of the transit: "[T]he Rev. Mr. Barton of Lancaster, who assisted me at the Telescope, on receiving my signal, as had been agreed, instantaneously communicated it to the counters at the window, by waving a handkerchief, who walking softly to the clock counting seconds as they went along, noted down their times separately, agreeing to the same second…"[209]

A newspaper notice informed the public about one of David's famed orreries (Figure 4.7): "[T]he Orrery, of which the American Philosophical Society formerly published an Account, projected and executed by Mr. David Rittenhouse, in this Province, is now almost finished. As this is an American Production, and much more complete than any Thing of the Kind ever made in Europe, it must give great Pleasure to every Lover of his Country…."[210]

Although war was raging at the time and David Rittenhouse was otherwise occupied, Thomas Jefferson wrote to him in 1778, requesting an

**Figure 4.6** Pine-case astronomical clock by David Rittenhouse. APS Collection 58.23. Photo by Brent Wahl.

**Figure 4.7** Orrery by David Rittenhouse, built 1770–1771 for the College of Philadelphia, case by Philadelphia cabinetmakers John Folwell and Parnell Gibbs. Courtesy of the University of Pennsylvania Libraries.

astronomical clock, and recalled: "[A] kind promise of making me an accurate clock, which, being intended for astronomical purposes only, I would have divested of all apparatus for striking, or for any other purpose, which, by increasing its complication, might disturb its accuracy. A companion to it for keeping seconds, and which might be moved easily, would greatly add to its value."[211]

Rittenhouse never fulfilled the commission. Eventually in 1811, Jefferson ordered from clockmaker and retailer Thomas Voigt a high-accuracy timepiece that now stands in the cabinet room at Monticello.

Thomas Jefferson had a letter from Francis Hopkinson describing a novel water clock. Hopkinson had a keen interest in timekeeping which he no doubt discussed with Edward:

I have sent in to the philosophical Society a Contrivance for the perfect Measurement of Time, and I see no Reason in Theory why it should not answer. I am now making Experiments to ascertain the Fact. My Device consists of a small glass Syphon, the shorter Leg of which is fixed in a Float of Cork or light Wood, which is to rest on the Surface of a Small Bason of Spirit, the longer Leg of the Syphon to come over the Edge of the Bason and to be drawn out to an exceeding fine Point so that the Liquor may fall in Drops only. Directly under the discharging Leg of the Syphon is a long glass Tube, such as one used for Barometers, which receives the falling Drops, and the Rise of the Liquor in the Tube is to designate the Hours and Parts of an Hour on a Scale pasted along Side of it for that Purpose. As the receiving Orifice of the Syphon must always be at the same Distance below the Surface of the Liquor, it must always be prest by the same Weight, and my

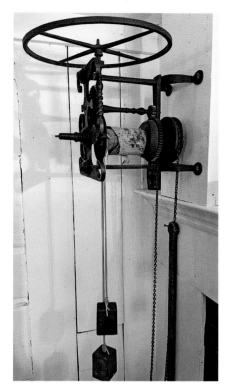

**Figure 4.8** Roasting jack at Hancock-Clarke House, Lexington, Massachusetts. Author photo.

Expectation is that equal Quantities of the Fluid will be discharged in equal Times.[212]

In Philadelphia, during the Spring of 1794, "Citizen SANS CULOTTE" and "Mr. L'ARISTOCRAT" exhibited:

[T]wo artificial men, who are of the ordinary size of man, perform feats of dexterity and surpass nature itself: they are set at one end of a room, entirely by themselves, having not the least connection with any thing, nor any person near them. when they are so to be put in motion, a person that attends does no more than winds up the moving springs that are contained within their bodies, they then, as it were receive life by degrees, salute the company and seem to rival each other to please the spectators with their agility. Their motions are formed to music, and they dance to many airs.... These curious automatons, or artificial men, after shewing many great feats of their activity and of lofty tumbling over an iron bar fixed horizontally, finish their labors in leaping a somerset backward and forward, and saluting the company...[213]

George Washington enjoyed the performance of these mechanical men that would have appealed to Edward as well. There were other public exhibitions of automata in Philadelphia in 1783, 1786, and 1796.

Finally, clockmakers also sometimes made other kinds of mechanical household instruments such as the rotating spits used in kitchens of the time. Clarke's Inn in 1745 had bowlegged "spit dogs" that were rounded up at cooking times to run inside hollow squirrel-cage wheels used for turning meat in kitchen hearths.[214] These were alternatives to singed servants hand-cranking the spits. Weight-driven clockwork roasting jacks, and smoke jacks powered by fan blades inside a chimney, were in use, too (Figure 4.8). In 1769 Alexander Smith advertised: "White-Smith, late of London, in Second-street, at the sign of the Lock Jack and Bell.... makes, mends and cleans all sort of jacks.... as yet been imperfectly performed in this place."[215] John Cadwalader, just one block away, purchased two jacks from Smith, who also installed a system of bells to summon servants in the grand homes of Cadwalader and others in the upper classes. Although jacks operated by gears and weights, they often were made by artisans like Smith who were not clockmakers. A jack and spit were listed in Edward's 1803 estate inventory but none made by him are known.

## POSTSCRIPT

The Thomas Stretch State House clock, formerly under Edward's care, was relocated in 1829 to the nearby steeple of St. Augustine's Catholic Church. Anti-Catholic rioters burned the church in 1844 and destroyed everything in it. Witnesses reported that the clock's bell pealed ten o'clock for the final time as flames engulfed the belfry.

INQUISITION indented and taken at *Philadelphia*, in the County of Philadelphia, the *Twenty Eighth* Day of *February* in the Year of our LORD One Thousand Seven Hundred and *Twenty two* before me JUDAH FOULKE, Esquire, High-Sheriff of the City and County of *Philadelphia*, by Virtue of the Writ of our Lord the King that now is, to me directed, and to this Inquisition annexed by the Oaths *and Solemn Affirmations* of *John Biddle, Joseph Redman, Edward Duffield, William Morrell, Peter Ruve, James Wharton, William Jones, Reynold Keen, Robert Erwine, James Craige, Joseph Bullock and William Smith*

good and lawful Men of the County aforesaid, who, upon their Oaths and Affirmations aforesaid, respectively do say, That the Rents, Issues, and Profits of *One full equal and undivided moiety or half part of and in a certain Two story Frame Tenement and Lot of Ground situate in the City of Philadelphia on the West side of Front Street containing in breadth Twenty feet and in depth about One hundred feet Bounded by said Front Street, Dock Street and Ground of Jedediah Snowden John Leacock and others*

by me the said Sheriff taken in Execution, by Virtue of the said Writ wherein *John Godby* is Plaintiff and *Emanuel Rouse* is Defendant, are *Not* of a clear yearly Value, beyond all Reprises, sufficient, within the Space of seven Years, to satisfy the Debt and Damages in the Writ, aforesaid mentioned. IN TESTIMONY WHEREOF, as well I the said Sheriff, as the Inquest aforesaid, to this Inquisition have interchangeably set our Hands and Seals, the Day and Year first above written.

*Judah Foulke* Sheriff

*Wm Jones*

*John Biddle*

*Jas. Wharton*

*Jos. Bullock*

*Reynold Keen*

*Jos Redman*

*Craig*

*Edwd Duffield*

*Rob Erwin*

*Wm Morrell*

*Wm Smith*

*Peter Reeve*

CHAPTER FIVE

# Edward Duffield: Citizen and Anglican

*March 26—Dined at Dunwoody's with Governor Mifflin, Benjamin Chew, Judge McKean, Edward Shippen, Richard Peters, General Wayne, Daniel Brodhead, Edward Duffield, Mayor Clarkson, Charles Jarvis, Capt. Anthony, William Jones, R. Keen, Tench Francis, Andrew Tybout, Judge Biddle and Joseph Donaldson.*

— Diary of Jacob Hiltzheimer, 1796

Far more than an artisan horologist, Edward had major civic, legal, and church responsibilities. He joined boards, juries, and religious and educational institutions, including the APS. These aspects of Edward's life rank equally with his horological endeavors, and they began just a year after he reached maturity. Clockmakers often were active citizens, but Edward's status as an affluent property-owner elevated him to higher levels of city and church engagement. Long-term friendship with the older Benjamin Franklin enhanced his reputation, as did his sister's marriage to Ebenezer Kinnersley. Edward's other connections included attachments with city luminaries such as the Duchés, Drinkers, Rittenhouses, and Hiltzheimers. With his friend Francis Hopkinson, he supervised the Bray School for Black children, and later in life he helped found and support schools near his Benfield estate.

**Figure 5.1** Signature on the 1752 original Deed of Settlement, courtesy of the Philadelphia Contributionship for the Insurance of Houses from Loss by Fire.

Edward came of age in 1751 and began to play a role in Philadelphia's public life almost immediately. In 1752, he signed the original roll of subscribers for America's first successful fire insurance association, the Philadelphia Contributionship for the Insurance of Houses from Loss by Fire, founded by Franklin and still in business today (Figure 5.1). A 1758 inspection by this association recording the details of Edward's home is still available (see Figure 8.2).

In 1755, he became a Christ Church vestryman, a prestigious lay position in the city's elite church. He held that office until 1772.

In 1756, Edward joined subscribers of a newly reprinted book, *A Voyage to the South-Sea in the Years 1740–1* written by John Bulkeley and John Cummins, crewmen of the *H.M.S. Wager*.[216] Bulkeley had fled from Britain to Pennsylvania after years of defending his actions at home following *H.M.S. Wager*'s shipwreck and the subsequent starvations, deaths, and miraculous return of a few survivors

1772 document with Edward Duffield's signature and seal. Author's collection.

from the desolate coast of southernmost South America. Edward possibly met Bulkeley, or heard him speak, and was inspired to subscribe to the book's publication because of this. The subscription proposal was advertised a few times in the *Pennsylvania Gazette*,[217] and the book was ready for subscriber pickup as announced on March 10 of the following year.[218] James Chattin, one of the city's prominent publishers, signed a receipt for Edward (Figure 5.2).[219]

Proof of the trust placed in Edward by his fellow citizens, in 1757 he was one of just twenty-four men entrusted to countersign 18,000 newly-printed paper banknotes issued by the colony.

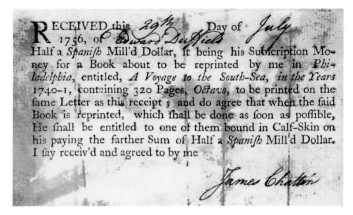

**Figure 5.2** Receipt to Edward Duffield for book subscription, Heritage Auctions, November 17, 2017, Lot 92005. Courtesy of Heritage Auctions, HA.com.

Edward was elected to the Young Junto in 1759. It was a revival of Benjamin Franklin's original Junto, founded a few decades previously, composed of Franklin and a small group of his close friends. Both clubs met regularly to discuss moral, political and philosophical issues, and a list of questions discussed still exists today.[220] There is no record of Edward attending and he did not sign the bylaws. Nor was he a member when the group reorganized two years later, so it is uncertain whether he ever actively participated in the club's activities despite his election.

Because of fears about public morality in 1761, Edward joined Kinnersley and many others in signing a public letter opposing lottery-funded public baths.[221] That was an unfortunate setback for the city's public hygiene. On September 4 of that same year, Edward participated as a vestryman in a procession from Christ Church to its sister church, St. Peter's, and in 1763 he became a warden of Christ Church, an even greater honor, along with William Plumstead (or Plumsted, 1708–1765) who was city mayor in 1750, 1754, and 1755 and a longtime friend of Franklin's. At one point, the two wardens were tasked to "collect subscriptions to help frontier inhabitants who had been obliged to fly from their habitations to avoid the Indian Ravages and cruel Murders."[222] This proved difficult since there were 790 desperate families needing assistance. However, Edward in 1766 was appointed as Treasurer to receive and distribute the funds, and a church committee of Joseph Redman and Joseph Swift later audited his work and reported that he had distributed £947 for which vouchers were examined.[223] Edward himself kept records of such church collections as noted in an 1841 book by Reverend Dorr, *Historical Account of Christ Church*. Dorr wrote on page 139 that the account book "was kindly lent me by E.D. Ingraham," Edward's grandson, but it is unlocated today.

This was only the first of Edward's charitable efforts in this period. A July 5, 1764, list of contributors to the Pennsylvania Hospital included Edward, who added £5 to his previous donation.[224] He again was a Christ Church warden, this year with Charles Stedman (1713–1784) who commissioned the Powel House that today is an historic house museum. A January 31, 1765, newspaper notice urged more donations to the poor due to the severity of the winter weather, and Edward was among those named to receive the contributions.[225] In 1776 with fellow church warden Charles Stedman, he petitioned the legislature to authorize a lottery to benefit St. Peter's Church to help repay large debts incurred in its construction.[226] Edward served as one of eight lottery managers promising 4,821 cash prizes.[227]

Records from the 1760s state that Edward held two seats in pew number 92 in Christ Church, close by the Franklins' three seats in number 59.[228] Later he paid yearly £1.10 rentals for his Christ Church main-floor, four-person Pew Number 76, also near seats held by Franklin's family, Robert Morris, and later by George and Martha Washington. Edward's pew rental continued for several years until he began worshiping at All Saints' Church, near Benfield after his move there.[229]

One of just two known letters written to Edward was sent by Reverend Richard Peters (1704–1776) from England in 1765 (Figure 5.3). Peters traveled there to obtain a preaching license from the Bishop of London. The letter is at the Library of Congress and is pasted inside a book which belonged to Edward's grandson Edward Duffield Ingraham, an avid book collector. Along with Edward's long-time service to his church, the letter suggested that he was a deeply religious man and indicated that Edward and Peters had a very close relationship:

To Mr. Edward Duffield, Watch Maker in Philadelphia, Liverpool 18 August 1765,

Dear Sir,

I have the pleasure to tell you that everything has been favorable to me since I parted with you at Chester. The winds were moderate & fair —no sickness —indeed I wanted no more purifications. My appetite came to be very good. My health grew better every day. We were but 32 days on the sea. On Saturday the 21 July I returned (I think that very day of the month I went from here) after an absence of twenty nine years to the embraces of my dear brother and sister. They have a greatly well educated family & live very comfortably. I mention these little things to you, as you have an affectionate temper & just such another partner and family. My love to them. You have distinguished me in so friendly and open a manner since I got out of the embargo of business that I owe you every thing that love can do where it meets with love.

The impressions that your great goodness brought upon my head in restoring me to the place I had in the Church, after so long an abdication of his service, brought me to a determination of giving myself up entirely to his divine Providence and Conduct. I have not for some time past desired, and I hope I never shall desire anything respecting myself either as to life or death, sickness or health, but just what God shows me in the occurrences of the present day. These are all of his appointments—these are all the real effects of his government of the world— and therefore all these are to be received with cheerfulness and to be used to his glory & the good of those with whom we are connected.

After satisfying the demands of my private affections I shall most certainly settle all matters with my relations so as to take a final farewell, if God does not take me to himself whilst I am among them.

They agree with me that religion calls for some distinctions that perhaps humanity would not bear to hear of—our family is to be preferred to our own selves—our country to our family—God is to be preferred to self—Family & County so this distinction which, thank God, I draw from the sacred truth taught by Christ Jesus—the remainder of my life is consecrated to the service of God and his churches in Philadelphia. Let all my dear connections both in Christ Church & Saint Peters know this, and then we shall be present together in spirit even while we are apart in body. I desire them & your prayers for me and am

Dear Sir, your affectionate humble servant, Richard Peters, 1765, September 19.

Further confirmation of the closeness of these two men is found in the minutes of Christ Church for June 1764. Edward joined fellow warden Charles Stedman in signing a letter to Peters prior to his departure for England (Figure 5.4). Edward's name and signature appear more than one hundred times during those twenty-three years as he attended meetings, handled finances, and signed letters and reports. The minutes for this period are both photographed and transcribed on the church website.[230]

The years 1766 and 1767 were busy years for Edward on other fronts. He was named as one of twelve directors of the Contributionship, and records show that he was present for nearly every monthly meeting. April 16 minutes placed Edward in the room with James Pemberton, Joseph Fox, Thomas Wharton, Samuel Morris,

**Figure 5.3** Reverend Richard Peters portrait by Mason Chamberlin, National Portrait Gallery, Smithsonian Institution, NPG.82.146.

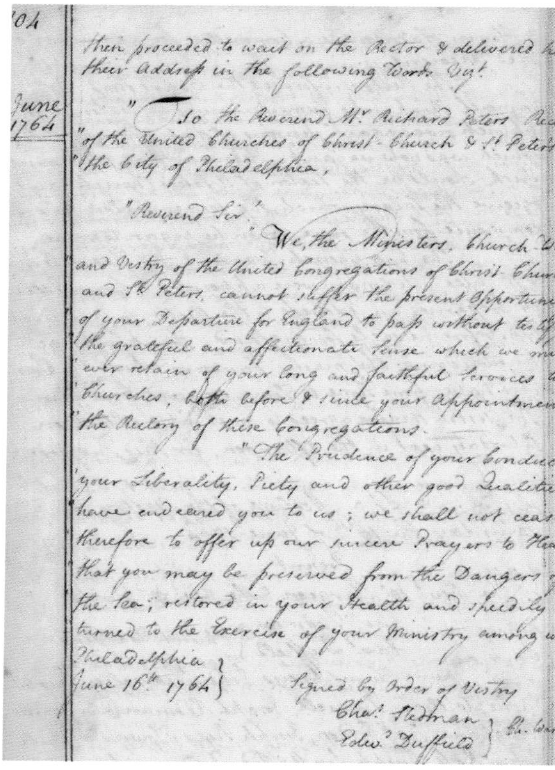

'04

then proceeded to wait on the Rector & delivered him their Addreſs in the following Words Viz:

June 1764

"To the Reverend Mr. Richard Peters Rector "of the United Churches of Christ-Church & St. Peters in "the City of Philadelphia,

"Reverend Sir,

"We, the Ministers, Church-Wardens and Vestry of the United Congregations of Christ-Church and St. Peters, cannot suffer the present Opportunity of your Departure for England to paſs without testifying the grateful and affectionate Sense which we must ever retain of your long and faithful Services to our Churches, both before & since your Appointment to the Rectory of these Congregations.

"The Prudence of your Conduct, your Liberality, Piety and other good Qualities have endeared you to us; we shall not cease, therefore to offer up our sincere Prayers to Heaven that you may be preserved from the Dangers of the Sea; restored in your Health and speedily returned to the Exercise of your Ministry among us."

Philadelphia June 16th 1764

Signed by Order of Vestry
Chas. Stedman
Edwd. Duffield
Ch. Wardens

Figure 5.4 Copy of letter to Peters in Christ Church Minutes. Courtesy of Christ Church

and brothers Thomas and Joseph Stretch. Edward also assisted in appraising the estate of Lewis Evans, a neighbor and a renowned surveyor and mapmaker. An Arch Street onsite auction of Evans' possessions was advertised in 1759 and included "Mathematical Instruments: Among which are a handsome Clock, with a Mahogany case,"[231] and we can imagine Edward having been tempted by the clock and surveying materials. Another neighbor, schoolmaster Joseph Stiles, opened classes a few doors from Edward's shop. He was joined the next year by William Ranstead who taught surveying and instrument-making, perhaps instructing Edward in these disciplines.

Edward unsuccessfully sought the post of Collector of the Excise for the City and County of Philadelphia in the same year. In January of 1767, he put himself forward to head a proposed office for assaying and stamping locally made gold and silver products. Modeled on the English system of testing and hallmarking, the post required substantial knowledge, skill, and trustworthiness. Local goldsmiths and silversmiths supported either him or a Quaker competitor, Owen Biddle. In the end the Assembly tabled both applications and did not establish the office.

In May of 1767, the Bray Associates had a report from Edward and Francis Hopkinson, both selected by Franklin to oversee the education of Black children in Philadelphia, about a planned school building.[232] A subsequent letter from Franklin urged the two appointees to proceed, with promised funds forthcoming, to purchase land.[233] The following January, Franklin wrote about the school to Hopkinson who replied in March with an update on progress being made.[234] More details about the Bray School are provided in Chapter 9.

On March 8, 1768, Edward was elected to the APS, further proof of his stature within the intellectual community. He joined two committees, Mechanics & Architecture and Husbandry & American Improvements, and he attended his first meeting on March 22. Other new members were his cousin Samuel Duffield, William Franklin (Benjamin's son), Joseph Fox, Richard Peters, Thomas Bond, and British General Thomas Gage. In the following year, the APS merged with its rival local society and named Franklin as president in absentia. They selected their Vice President Thomas Bond to manage the united group. Edward served on special committees addressing the manufacture of paper and inspecting a model of a new dredging machine.

Edward was one of twenty-two citizens appointed to serve under Sheriff Joseph Redman during the 1770 April term of "A Grand Inquest," a grand jury that handed down indictments of accused criminals. He served again in July, 1770. Edward also contributed money to various community efforts in that year. He subscribed £2 for promoting the local production of silk. This domestic-industry effort received considerable support but ultimately failed. He also contributed to a £6,000 city payment to the King and was an Assistant Judge in the process.

Edward's civic role continued to grow and in 1771 he was appointed a city warden overseeing the night watch, whose duties included calling out the time and weather each hour, and maintaining whale-oil street lamps and water pumps. A special tax was levied to support those public services. As the population expanded, there was a shortage of freshwater pumps in some neighborhoods, making this a vital issue for the wardens. In October of the same year he was appointed by the Provincial Council to a committee considering a new road. Fellow committeemen were Joseph Fox, Jacob Lewis, and Jonathan Evans. They issued a favorable report the following April. He was also asked to be a director of the Library Company of Philadelphia, having purchased a share three years earlier, but he declined to serve "as his business would not permit."[235] He later bequeathed the share to his son Edward, Jr.

In 1772 Edward was one of four city wardens enforcing designated standards and rules for city draymen, wagoners, carters and porters.[236] He continued his financial generosity as well by donating £5 to the College of Philadelphia which later became the University of Pennsylvania.

In October 1773, Edward became a Philadelphia County Commissioner,[237] but much of his attention seems to have been focused on the Bray Associates' project for Black children. He and Francis Hopkinson informed the Bray Associates that after careful inquiry they had located a suitable lot on which to build a school.[238] It had a 136-foot frontage on Market Street and extended 360 feet along Ninth Street. With Franklin present at a September meeting of the Associates in London, the purchase was approved for a price of £575.

Although most of the Associates correspondence was written by Hopkinson, a May 1774 letter from Edward to Waring was discussed at a September meeting. Edward offered data on the students. This is just one of two known letters on any subject penned by Edward:

> Sir, I have drawn on you of this date for Ten pounds Sterling in favour of Mr. Thomas Barton or Order, being for half a years Salary due the Mistress of the Negroe Charity School in this City, which I doubt not you will duly honour. Mr. Hopkinson (who resides in the Country) not being in Town at this time occasions my signing this Draught alone. Mr. Coombe hath not as yet joined us in any of the Associates business. I am Sir your most Obedient Humble Servant Edwd. Duffield.
>
> N.B. It appears by the Account of the Negroe Charity School rendered this day by the Mistress thereof, that there are at present 2 at their Needles and Spelling, 1 at Knitting,

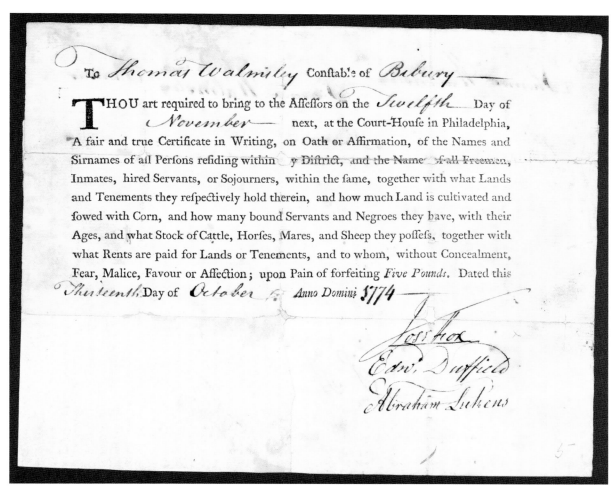

**Figure 5.5** October 13, 1774, constable's warrant signed by Edward Duffield. Collection of the Historical Society of Pennsylvania, Miscellaneous Collection (0425) Box 5B.

**Figure 5.6** 1799 print of Walnut Street Goal by W. Birch & Son. Courtesy of the Library Company of Philadelphia.

Needle, & Testament, 7 at Spelling, 3 in the Testament, 1 in the Psalter, 10 in the Alphabet, 1 in Fables, 3 in the Primer, Three of the above Children are free & the rest of them are slaves.[239]

That May Edward and Hopkinson informed the Associates that the latter had already departed from Philadelphia, and that Edward "intended very soon to retire into the County, & therefor requested that some other person might be appointed Inspector of the Negroe School in their Stead."[240] Thomas Coombe was appointed but with the approaching political turmoil, the Bray School soon foundered until being revived in December 1786, and continuing until 1845.

On July 29 of 1774, Edward joined his fellow county tax commissioners, Abraham Lukens and Joseph Fox, at a meeting of tax collectors ordered to appear with records and funds (Figure 5.5). The commissioners also met in several previous months with their local assessors. In November, a warrant was signed by Edward, Lukens, and Fox directing Constable Thomas Walmsley of Bibury to provide them with a written certificate naming every resident and their properties.[241]

Edward also joined Fox as an inspector and manager of the construction of the innovative Walnut Street Jail at the corner of Sixth Street (Figure 5.6). The brick two-story, two-wing, fireproof structure, 32 by 184 feet, was designed and built by Robert Smith as a model penal institution that was studied by Colonial and overseas visitors. When it was completed in September, Edward contributed to the customary dinner for the workers and supporters. Jacob Hiltzheimer reported in his diary that he attended the event.

Sometime shortly thereafter, Edward moved to Benfield and records of his civic involvement in the city of Philadelphia end, although he later held county positions and supported local schools near Benfield. In fact, the only account of his activities between the end of 1774 and 1777 is an advertisement offering a twenty-shilling reward for a runaway Scottish servant girl, Anne Munro, who had absconded from Benfield (Figure 5.7).

In July 1777 we hear of Edward again when he paid a £22.10 fine for non-performance of militia duty in Philadelphia County's Third Class Company led by Captain Samuel Swift, Edward's brother-in-law. The fines ranged up to £30.8, rose to £40 the following year, and were £100 by 1779. The money paid for substitutes, per a new Pennsylvania law that made military service compulsory for free men aged eighteen to fifty-three.

Had Edward been in uniform, he might have fought in the October 4 Battle of Germantown where the militia, egged on by Continental regulars behind them, was heavily engaged, but forced to retreat when the regulars failed to support them. An Edward Duffield is listed as a surgeon's mate in Colonel William Bradford's battalion, but this seemingly was a younger cousin who served under another cousin, surgeon Samuel Duffield. Edward's son, Edward, Jr., was too young to fight.

RUN away on the 19th of May laſt, a ſervant GIRL, named *Anne Munro*, late from Scotland, aged about 18 years, low of ſtature, fair complexion, brown hair, and ſuppoſed to be with child; ſhe had on, when ſhe went away, a green baize ſhort gown, ſtriped lincey petticoat, tow apron, and check handkerchief. Whoever will ſecure the ſaid *Anne Munro* in the Goal or Work-houſe in Philadelphia, or bring her to the ſubſcriber, in the Manor of Moreland, Philadelphia county, ſhall be paid TWENTY SHILLINGS, with reaſonable charges, by
*Benfield, July* 16, 1776. ¶ EDWARD DUFFIELD.

**Figure 5.7** Edward Duffield advertisement seeking runaway servant girl, *Pennsylvania Gazette*, July 24, 1776, p.1. Courtesy of Newspapers.com.

PHILADELPHIA, Feb. 18.

On the 4th inſt. Captain Hoveden, with a part of the Philadelphia troop of light dragoons, paſſed up the Briſtol road, and Capt. Thomas, with a part of his independent company of Bucks county volunteers, paſſed up the Buſtletown road, with orders to take a number of militia officers, juſtices of the peace, and collectors of fines, under the new ſtate, and ſome other diſaffected perſons: On their return Capt. Hoveden brought in with him the following priſoners:

Gunning Bedford, former commiſſary of muſters, and now detected of robbing the market-people, in company with two more.

Edward Duffield, on ſuſpicion of being active againſt governmet.

**Figure 5.8** Newspaper report of Edward's capture by British dragoons, *New-York Gazette, and Weekly Mercury*, March 23, 1778, p.2. Courtesy, American Antiquarian Society, accessed via GenealogyBank

In February 1778, during the British occupation of Philadelphia, a party of twenty-four British light dragoons led by Captain Hoveden skirmished into Bucks County. They engaged with Continental troops, seized loads of cloth meant for Washington's army, and returned with a number of county leaders including Edward. Ironically, the prisoners were confined in the Walnut Street jail that Edward recently helped to build.

A Philadelphia Loyalist newspaper praised the mission and listed the men captured, but Edward's name oddly was not included.[242] He did appear in an account in a New York Loyalist paper (Figure 5.8), which stated that he was detained "on suspicion of being active against government."[243]

Joseph Kirkbride (1731–1803) wrote to the Governor about Edward's capture: "We have been allarm'd by the Enemy's coming out into the Neighborhood of the Red Lyon, & have Captur'd some more of our Friends, particularly Mr. Jno Vandergrift, a Commissioner for this County, & his Son, Mr. Edward Duffield, & several others not yet known to me."[244]

Kirkbride was a Bucks County farmer, local official, and member of the APS together with Edward. He added, "The Enemy having lately Burnt two Valuable Dwelling Houses with all my Out Houses of Every kind & sort, & a great deal of

Furniture, Utensils, Corn, Hay &c, & Intirely Dislodging my Family…"[245] Edward was fortunate not to have suffered similar destruction at Benfield. Thomas Paine wrote to Franklin that he spent two nights at "Mr. Duffels in the winter… Mr. Duffels has since been taken by them and carried into the City but is now in his own house."[246]

Paine's letter, and Edward's name being omitted from the Philadelphia paper, suggest that Edward was arrested but quickly released. This may have been due to his connection with Loyalist publisher James Humphreys, Jr, with whom Edward had cosigned a public letter in 1761. Edward's mother-in-law's maiden name was Humphreys, indicating that there might have been a family bond.

In September of the same year, Edward was an executor of the estate of Andrew Duché. Edward had many deep connections with the Duché family including the marriage of his aunt Esther to Colonel Jacob Duché Sr., a city councilman, mayor, and Christ Church vestryman.

An entry in Jacob Hiltzheimer's diary for January 3, 1783, recorded that Edward was a witness against tax collectors who two years previously had demanded excess payments for a tax supporting Continental Army enlistments. The Grand Jury found in favor of the complainants and issued a true bill.

Edward was elected in 1785 to the newly formed Philadelphia Society for the Promotion of Agriculture and Domestic Manufactures that then met regularly in Carpenters Hall. Still in existence, the Society states on its website:

> The Philadelphia Society for Promoting Agriculture is the oldest agricultural society in the United States. It has played key roles in developing many of the farming methods and institutions responsible for the abundance that is the hallmark of our modern food system. Practices now commonplace—such as crop rotation and management of soil fertility—have their roots in the robust dialogue and exploration of change that have marked the Society's meetings and publication activities from its origins to the present day.[247]

Edward's election to the Society is another confirmation that after his move to Benfield he had turned his attention to agriculture rather than clockmaking.

In the minutes of a March 6, 1787 meeting, Edward was thanked for making a drill, attached to a plough, for seed planting. Such seed drills were a literally groundbreaking 1701 invention by Jethro Tull in England, replacing inefficient and slower hand-scattering or planting of seeds one by one, and would have been a valuable item for the Society. Edward's making of the drill demonstrates his abilities in the crafting of specialized metal tools. The minutes from a meeting on January 11, 1791, note that Edward was one of the men present at another Society meeting. Elite attendees at this meeting included Powell (president), Hiltzheimer, Bache, and Morris.[248] Former President Washington, farming again at Mount Vernon, became a Society member at the same time.

In 1785, Edward was an election judge for the First District of Philadelphia County and again for the Second District, certifying election tallies and results. In the same year he appeared in Jacob Hiltzheimer's diary on September 18: "Returning from my visit to Trenton, I called at Mr. Edward Duffield's. He was absent from home, but Mrs. Duffield entertained me kindly with dinner and wine."[249]

From September 27 to October 8 of that year, Edward was a delegate to the first multi-state General Convention of the Protestant Episcopal Church. Convened in Philadelphia, it was preceded by several state and regional conventions. Edward was not noted as active in the deliberations. Edward's friends and associates in attendance included Richard Peters, Samuel Powell, Edward Shippen, and clockmaker John Wood, Jr. The participants approved a revised prayerbook published the following year.

Edward's name appeared several times in Jacob Hiltzheimer's diary. In October 1786, Hiltzheimer observed legal proceedings as a "struck juryman" at the State House and then "Took Mr. Edward Duffield home with me, who was one of the judges."[250] He breakfasted with Edward the next morning and returned to the State House with him.[251] The next year, the diarist recorded in February: "my wife and I went in my sleigh to Mr. Edward Duffield's; overtook Mrs. Duffield and her son Edward four miles from town; took her in with us, one of her horses being lame… About 12 o'clock it began to rain, which made us leave Mr. Duffield's."[252] Finally, in March of 1790, Hiltzheimer notes that he was visited by "Mr. Duffield and wife"[253] following the death of Hiltzheimer's wife, who died after a long illness and was buried in the Friends' graveyard at Fourth and Arch streets.

In 1789, Edward still was a shareholder of the Library Company of Philadelphia that published a new catalogue of its books. A few of those focused upon horology: *The principles of Mr. Harrison's time-keeper* (London, 1767); *Elements of clock and watch, adapted to practices*, with plates (London, 1766); and *Ferguson's dissertation upon the phaenomena of the Harvest-moon, also, the description and use of a new four-wheeled orrery, and an essay upon the moon's turning round her axis* (London, 1747). Edward perhaps borrowed and perused those volumes.

In May of 1791, Edward's circle of friends and acquaintances suffered another loss. Francis Hopkinson died suddenly of a stroke and was buried at Christ Church. Records do not indicate if Edward was present at the funeral.

Having moved to Benfield, Edward became more active in the smaller community there. In 1793, he joined his neighbors John Swift, William Walton, and Isaac Comly at a meeting about the repair of the road between Byberry and Moreland. They decided that the costs of the project would be divided between the two towns. The next year, he became deeply involved in construction of the Lower Dublin Academy to replace an old log schoolhouse. He was asked to draw up a plan for the new building and his large exterior clock was installed in its facade, as described previously. In 2016, local historian Joseph J. Menkevich prepared an application seeking historic designation for the decaying structure. He described the charter, planning, and construction which, due to several long delays, was not completed until 1803. Menkevich included details from an 1802 report:

> This Charter was signed by the Governor the 23rd of January 1794. On the receipt of it the board met in the old School house, and taking into consideration the smallness of the building, it was proposed and seconded that an immediate subscription should be opened to enable them to build a larger and more convenient building for the purposes of the Academy. This being done and a subscription started, Mr. Edward Duffield was requested to draw a plan for said building. The plan was approved of, the size of the house to be fifty feet long and thirty wide,—two stories high, the cellar to be seven (7) feet in the clear, the first story twelve (12) feet, and the second story eleven (11) feet. — the whole to be built of stone. At the same meeting Edward Duffield, Thomas Paul, and John B. Gilpin were chosen managers to superintend the building, appoint workmen, procure materials &c. Robert Lewis was appointed treasurer, and John Holme, Thomas Holme and Humphrey Watermen to collect subscriptions. During the year 1794, some material was collected, the well and cellar dug and the pump fixed, but from unavoidable causes the house was not yet built. However, the gentlemen appointed to collect subscriptions were requested to call on subscribers so there may be funds ready to go with the building early in the Spring of 1795.[254]

In 1795 Edmund Hogan published a city directory: *Prospect of Philadelphia, and Check on the Next Directory. Giving, at a single view, the numbers of the houses, names of the streets, lanes, courts, and alleys; with the names of the present inhabitants, and their occupations….* Edward lived outside the city and was not listed but his physician son Benjamin was among residents endorsing the book. Philadelphia historian Torben Jenk praised this source of information and shared Edward-related details unearthed within it:

> A little searching shows that Edward Duffield's clock shop on the northwest corner of Second and Mulberry/Arch streets was then occupied by John Redman, grocer (p.32). The former Duffield house at 110 High Street (now 320 Market) was occupied by Seth Craig, Saddler (p.10). Next door at 112 High Street was Benjamin Franklin Bache, Printer of the Aurora (and grandson of his namesake).[255]

Edward appeared in Hiltzheimer's diary on March 26, 1796, when Hiltzheimer recorded that he "Dined at Dunwoody's on Market Street…"[256] with a group that included Edward. Other prominent citizens at table were Governor Mifflin, Benjamin Chew, Judge McKean, Edward Shippen, Richard Peters, General Wayne, Tench Francis and Judge Biddle.

On August 26, Edward drafted a long letter to the younger Richard Peters in reply to formal queries regarding the use of "Plaister of Paris" or gypsum fertilizer. Along with his 1774 letter to the Bray Associates, this is Edward's only known letter. Peters was a judge of the United State District Court in Pennsylvania and first president of the Philadelphia Society for the Promotion of Agriculture and Domestic Manufacture. Edward's four-page letter was published in Peters' 1797 book printed in Philadelphia, *Agricultural Enquiries on Plaister of Paris. Also, Facts, Observations and Conjectures on that Substance, when Applied as Manure. Collected, chiefly from the practice of farmers in Pennsylvania, and published as much with a view to invite, as to give information*. Peters added that Edward's farm was an agricultural showplace, with a herd of tame deer and a cache of Indian arrowheads that were donated to local historian Pierre Du Simitiere. There was no mention of clockmaking.

Edward's letter provided many technical and scholarly details on his methods, studies, and crops. Regarding how long he tested the fertilizer:

> Every year since 1783… I repeated the plaister at four bushels to an acre…. plaistered five times since 1783…. Its effect is immediate upon grass of all kinds, and upon Indian corn, and upon all other kinds of grain the year following, when it is well mixed with the soil by ploughing…. the best time to apply the plaister is as soon as the barley or oat is taken off; as it gives a good growth to the clover… You will find by Dr. Berman, who has analyzed this fossil, that it contains twenty-two parts water, thirty-three parts calcareous earth, and forty-five parts vitriolic acid. And you also find in a small work by Dr. Home of Edinburgh, upon the principles of vegetation, a variety of accurate experiments continued for the space of four years….[257]

Peters sent the book to George Washington at Mount Vernon. The former President thanked him for the book's dedication and requested additional copies. Washington no doubt noted Edward's treatise. The men may well have been acquainted through many mutual friends including Peters. Washington often visited Judge Peters' grand home, Belmont, situated on a bluff overlooking the Schuylkill River.

In 1799, J.B. Bordley published his 600-page *Essays and Notes on Husbandry and Rural Affairs*, summarizing much that was known about agriculture. He printed excerpts from Judge Peters' book including Edward's work with gypsum.

In an 1889 biography of Benjamin Franklin by John T. Morse, Jr., the author wrote that Franklin many years earlier had promoted this fertilizer. "In a field by the roadside he wrote, with plaster, THIS HAS BEEN PLASTERED; and soon the brilliant green of the letters carried the lesson to every passer-by."

The last two references to Edward's participation in public life are from the first year of the new century. The first involved establishing the Maple Grove School in Powelton. Edward's neighbor

Aaron Walton asked Edward to donate land for another new local school and Edward agreed if Aaron would give half from his own land, which he did. The second is a notation that Edward was one of 156 subscribers to William Birch's *Views of Philadelphia*. Its twenty-nine prints were among the earliest depictions of the city. The bound and hand-colored copy cost $44.50, and is extremely rare and valuable today. A few of its plates are illustrations in this book. In his estate inventory three years later, Edward's copy was valued at only three dollars, perhaps because it was damaged, or the subscription price was mostly a donation to ensure the project's completion.

*Philadelphia's Old Courthouse, Town Hall, and Market in 1710 (Demolished 1830).*

" Edward Duffell to Jos Crisford
" Son Benjamin Coate and Brichis ......... 0" 17"
" Buckrom and Stais .......... 0" 5"
" tape ......... 0" 2"
" doon for Coate Lyning ......... 3/6 ... 0" 7"
" ron for Sleaves Lyning and Pockits ...... 8/2 ...
" oshed Buttons for Coate ......... 4/6 ... 0" 1 9"
" Brichis ......... 2/3 ... 0" 2"
" your Brichis ......... 0" 2"
" Silk Garters for Brichis ......... 0" 2"
" pair of Black Stocking Brichis ......... 0" 6"
" new Buckrom and Stais ......... 0" 2"
" tape and fusing ......... 0" 1"
" garters 2/6 and 18 Buttons ......... 8 1/ ... 0" 3"
" your Brichis ......... 0" 3"
" Son, Edward Benjamin for his Coate Brichis 0" 15"
" Buckrom and Stais ......... 0" 5"
" tape ......... 0" 1"
" Brown Hollon for Coate Sleaves & Pockt fair 2/8 0" 4"
" Coate 1/6 and 14 Buttons for Brichis ... 8d ... 0" 2"
" Brichis ......... 0" 2"
" for Son Benjam Coate & Waistcoat yours ......... 0" 18"
" Silk for Coate and Black Plush Waistcoat ......... 0" 2"
" rom and Stais ......... 0" 3"
" mohair for for Coate and Vest ......... 0" 2"
" and pockets for Vest ......... 0" 1"
" Coate ......... 1/6 ... 0" 2"
" Coate and Vest ......... 0" 4"

CHAPTER SIX

# Attire of Edward Duffield and his Fellow Philadelphians

*Know first who you are, and then adorn yourself accordingly.*

—Epictetus

*The Episcopalians showed most grandeur of dress and costumes… the arrival of the worshippers in damasks and brocades, velvet breeches and silk stockings, powdered hair and periwigs, was a sight to see.*

— Gottllieb Mittelberger, Mid-eighteenth-century visitor to Philadelphia

The ledger of upscale-tailor Joseph Graisbury itemized Edward's ninety-nine clothing purchases between 1765 and 1773. A complete transcription below shows prices for fine suits, vests, garters, coats, britches, and stockings for Edward, his two sons, and his "Negro man." Several fabrics were listed, along with hundreds of buttons, sixty-four of them silver-plated. No similar records have been discovered for dressmakers serving Edward's wife and daughters, but those women must have dressed stylishly as well.

Fifteen years before Edward's first entry in Graisbury's ledger, German immigrant Gottlieb Mittelberger arrived in Philadelphia in 1750 and later published his impressions of the city, its people, and its fashions. More from his book is in Chapter Seven. He described popular clothing and hair styles:

> Throughout Pennsylvania both men and women dress according to the English fashion. Women do not wear hoop-skirts, but everything they do wear is very fine, nice, and costly. Skirts and jackets are cut and sewn in one piece. Skirts can be parted in front. Under them women usually wear handsomely sewn petticoats trimmed with ribbon. But the outer long skirts have to reach down to the shoes, and are made of cotton, chintz, or other rich and beautiful material…. When the women walk or ride out, they wear blue or scarlet cloaks reaching down to the waist. On their heads they wear black or beautifully colored bonnets

*Detail of page from Joseph Graisbury's account book showing Edward Duffield's clothing purchases.*

instead of straw hats.... All Englishwomen are generally beautiful; they usually wear their hair cut short or trimmed....The apparel of the men, especially the English, is generally very elegant and this applies to farmers as well as to the other ranks. It is all made of excellent English cloth or similar material; and the shirts are all fine. Peasants as well as gentlemen wear wigs. In Philadelphia very large and very fine beaver hats are worn.... Everyone wears long trousers that reach down to the shoes; such trousers are very wide and are made of fine stiffened linen. All the men have their hair cut quite short during summers, and wear only a cap of fine white linen and over it a hat with the rim not turned up. On entering a house they doff the hat, but not the cap. And if anyone travels even just an hour's journey over land, he wears his long coat and a pair of boots that are half turned down and reach only to the middle of the calf.[258]

Later in the century, a 1787 letter from Philadelphia merchant John Siddon Whitall (1757–1843) offered advice to his nephew approaching his eighteenth birthday. John listed apparel the lad should own, similar to clothing that would have been worn by Edward's sons and younger relations:

> [P]iece of good Irish linen made up into shirts — with a cambrick Neck handkerchief to each shirt ….. 1 pair of Good Boots, 4 pair of good Shoes, 1 pair of buckles— *fashionable*—plates, 1 Great Coat—or Surtoot, 6 pair worsted stockings, 2 pair yarn ditto to ware under boots, 6 pair thred ditto fine, 2 pair ditto— coarse to ware under boots, 7 pair of linen drawers—1 pair of breeches, 2 waistcoats — *fashionable patterns* —fancy, 1 pair of spurs— *plated fashionable*, 1 horse whip, 1 Broadcloth coat—made in the fashion—with fashionable buttons—1 Round Hat—for riding—*Castor*, with the clouths you already have—will be sufficient to begin with.[259]

A 1782 portrait of prosperous Philadelphia merchant, politician, and bank president Thomas Willing (1731–1821) illustrates how a man of Edward's social position dressed at that time (Figure 6.1).

Another post-Revolution portrait by Peale in 1787 illustrated attire of an upper-class gentleman, Thomas McKean (1734–1817), and his son. McKean lived in

**Figure 6.1** Thomas Willing by Charles Willson Peale, 1782. Metropolitan Museum of Art, 66.46.

**Figure 6.2** Portrait of Chief Justice Thomas McKean and His Son, Thomas McKean, Jr., 1787 by Charles Willson Peale, Philadelphia Museum of Art, 1968-74-1, Bequest of Phebe Warren McKean Downs, 1968.

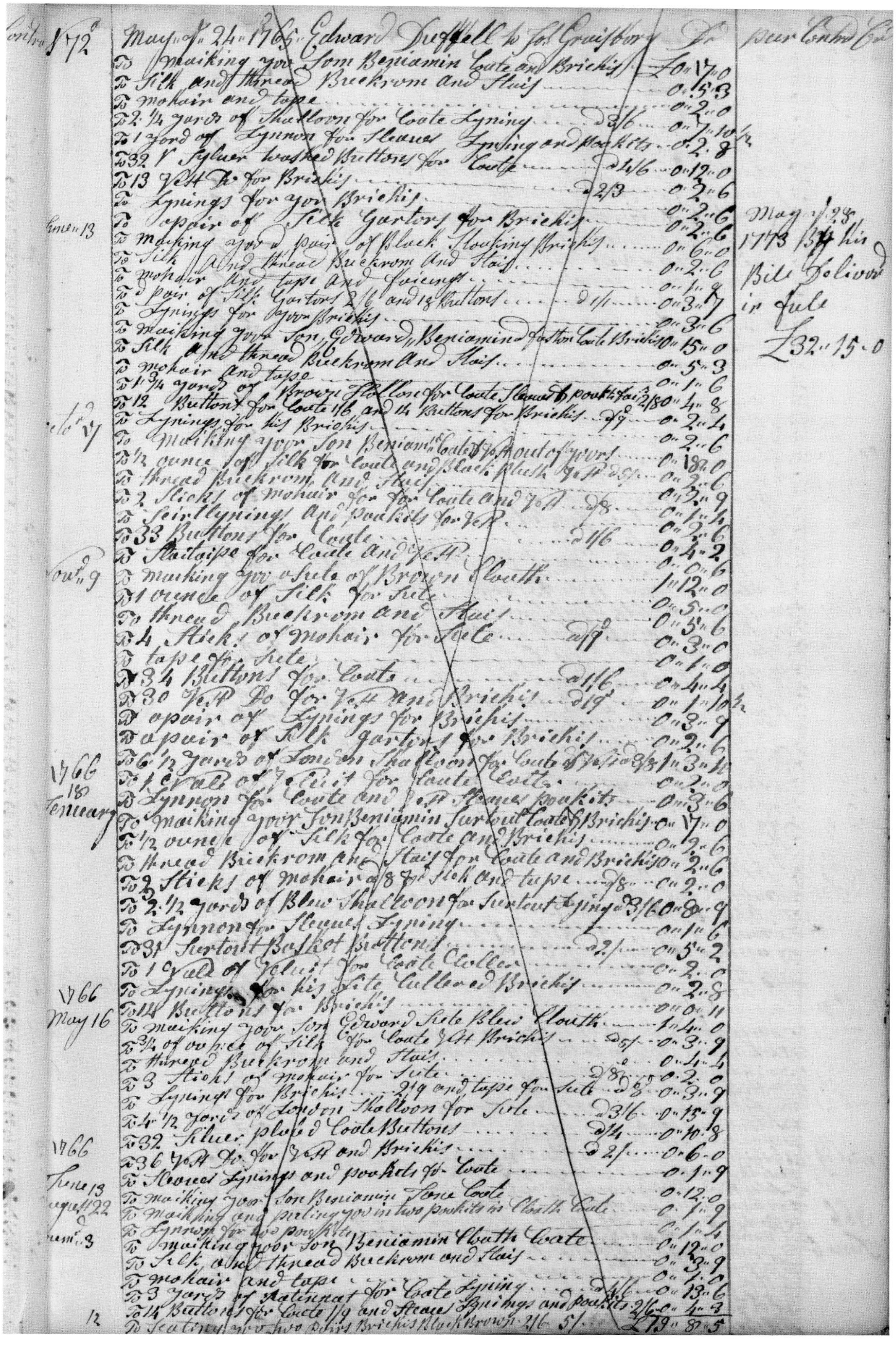

**Figure 6.3** Collection of the Historical Society of Pennsylvania, The Graisbury Ledger in the Reed and Forde Papers #501467, p.72. Photo by Keirstyn Allulis.

central Philadelphia and was deeply involved in politics as a delegate to the Continental Congress and as a Pennsylvania governor (Figure 6.2). Edward would no doubt have dressed similarly.

Graisbury's 130-page ledger is preserved at the Historical Society of Pennsylvania.[260] The first of Edward's two pages is shown in figure 6.3.

Joseph Graisbury was an unusually affluent Anglican tailor who, like Edward, had high-level connections. His shop produced bespoke elegant uniforms in 1775 for the militia and he was sufficiently prosperous to employ other tailors for the cutting and stitching. Clockmaker John Wood, Jr., was also a customer but made just ten modest purchases.

## TRANSCRIPTION OF JOSEPH GRAISBURY LEDGER ENTRIES OF SALES TO EDWARD DUFFIELD

Original capitalizations and spellings have been retained. Some fabric names are not familiar today. "Buckrom" or buckram is a stiff loose-weave cloth of cotton, linen, or horsehair. "Shalloon" is tightly woven wool. "Holland" cloths are any fine plain-woven linens. "Satinnat" or satinet is a thin silk satin. "Durant" was an English glazed plain-woven woolen material. "Pershon" most likely refers to "Persian," a thin silk used for linings. "Staitape" is "stay tape" which would have been used to stabilize a fabric edge or seam.

Thanks are due to Margaret Maxey, HSP researcher; to Keirstyn Allulis, who combed the ledger for relevant entries; and to Lynne Bassett, textile and costume historian, who kindly reviewed, explained, and corrected the transcription. In an email, Lynne highlighted the entry for "Black Nit Brichis," explaining that "there's a great example of black silk knit breeches in the collection of Old Sturbridge Village."

May of 24 1765 Edward Duffell to Jos Graisbury for purc?
(In right column) May ? 1773 by his bill ? In full £32.15.0

| | |
|---|---|
| To making your son Benjamin Coate and Brichis | 0.17.0 |
| To Silk and thread for Buckrom and Hair | 0.5.3 |
| To Mohair and tape | 0.2.0 |
| To 1/4 yard of Shalloon for Coate Lyning @3/6 | 0.8.10 1/2 |
| To 1 yard of Lynnon for Sleave Lyning and pockits | 0.2.8 |
| To 32 sylver washed Buttons for Coate @4/6 | 0.12.0 |
| To 13 Vest Do for Brichis. @2/3 | 0.2.6 |
| To a pair of silk gartors for Brichis | 0.2.6 |
| To Lynings for your Brichis | 0.2.6 |
| To Making you a pair of Black Stocking Brichis | 0.6.0 |
| To Silk and thread Buckrom and Hair | 0.2.6 |
| To mohair and tape and faicings | 0.1.9 |
| To pair of Silk Gartors 2/6 and 12 Buttons @1/1 | 0.3.7 |
| To Lynings for your brichis | 0.3.6 |
| To Maiking for Son Edward, Benjamin father Coate Brichis. | 0.15.0 |
| To Silk and thread Buckrom and Hair | 0.5.3 |
| To Mohair and tape | 0.1.6 |
| To 1 3/4 yard of Brown Holland for Coate | 0.4.8 |
| Sleaves & pockets facings | 2/8 |

| | |
|---|---|
| To 12 Buttons for Coate 1/6 and 14 Buttons for Brichis 2/0 | 0.2.4 |
| To Lynings for His Brichis | 0.2.6 |
| To Maiking your Son Benjamin Coate & Vest out of yours. | 0.18.0 |
| To 1/2 ounce of Silk for Coate and black plush vest @5/ | 0.2.6 |
| To thread Buckrom and Hair | 0.2.9 |
| To 2 Sticks of Mohair for son Coate and Vest 2/8 | 0.1.4 |
| To ? lyning and pockits for Vest | 0.2.6 |
| To 33 Buttons for Coate @1/6 | 0.4.2 |
| To Staitape for Coate and Vest | 0.0.6 |
| To Maiking you a Sute of Brown Cloath | 1.12.0 |
| To 1 ounce of Silk for Sute | 0.5.0 |
| To thread Buckrom and Hair | 0.5.6 |
| To 4 sticks of Mohair for Sute 2/9 | 0.3.0 |
| To tape for Sute | 0.1.0 |
| To 34 Buttons fo Coate @1/6 | 0.4.4 |
| To 30 vest Do for Vest and Brichis 2/9 | 0.1.10 |
| To a pair of Lynings for Brichis | 0.3.9 |
| To a pair of Silk Gartors for Brichis | 0.2.6 |

1766 ? January

| | |
|---|---|
| To 6 1/2 yards of London Shalloon for Coate & Vest @3/8. | 1.3.11 |
| To 1 ? of velvit for coate coller | 0.2.0 |
| To Lynnon for Coate and sett Sleave pockits | 0.3.6 |
| To Maiking your Son Benjamin Surtout & Coate & Brichis | 0.17.0 |
| To 1/2 ounce of Silk for Coate and Brichis | 0.9.6 |
| To thread Buckrom and Hair for Coate and Brichis | 0.2.6 |
| To 2 sticks of Mohair and 8 yard Silk and tape @2/8 | 0.2.0 |
| To 2 1/2 yards of Blew Shalloon for Surtout Lyning @3/6 | 0.8.9 |
| To Lynnon for Sleeves Lyning | 0.1.6 |
| To 31 Surtout Basket Buttons. @2/ | 0.5.2 |
| To 1 yard of Velvit For Coate Coller | 0.2.0 |

1766 May 16

| | |
|---|---|
| To Lyning for His Sute Collar & Brichis | 0.2.8 |
| To 14 Buttons for Brichis | 0.11 |
| To Maiking your Son Edward Sute Blew Clouth | 1.4.0 |
| To 3/4 of ounce of Silk for Coate Vest Brichis 5/ | 0.3.9 |
| To thread Buckrom and Hair | 0.4.4 |
| To 3 sticks of mohair for Sute | 0.2.0 |
| To Lyning for Brichis 2/9 and tape for Sute | 0.3.9 |
| To 4 1/2 yards of London Shalloon for Sute @3/6 | 0.15.9 |

1766 June 13, August 22

| | |
|---|---|
| To 32 Silver plated Coate Buttons. 2/4 | 0.10.8 |
| To 36 Vest Do for Vest and Brichis. @2/ | 0.6.0 |
| To Sleaves Lynings and pockits for coate | 0.1.9 |

| | |
|---|---|
| To Maiking your son Benjamin ? Coate | 0.12.0 |
| To Maiking and puting you in two pockits in Cloath Coate | 0.1.9 |
| To Lynnon for two pockits | 0.1.4 |
| To Maiking your Son Benjamin Cloath Coate | 0.12.0 |
| To Silk and thread Buckrom and Hair | 0.3.9 |
| To Mohair and tape | 0.1.0 |
| To 3 yards of Satinnat for Coate Lyning @4/6 | 0.13.6 |
| To 14 Buttons for Coate 1/9 and Sleave Lynings and pockets 2/6. | 0.4.9 |
| To Seating you two pairs Brichis | 19.8.5 |
| Black Brown | 2/6 5/ |

February of 12 1767 Edward Duffell to Jos. Graisbury for puc

| | |
|---|---|
| To your account from Page 72 | 19.8.5 |
| To Maiking your Negro man Blew Cloath Vest | 0.6.6 |
| To Silk and thread Buckrom and Hair | 0.3.0 |
| To mohair and tape | 0.1.0 |

20

| | |
|---|---|
| To Maiking your Son Benjamin Cloath Vest out of his Vest. | 0.5.0 |
| To Silk and thread Buckrom and Hair | 0.2.6 |
| To Mohair and tape | 0.1.0 |

1767 21 March

| | |
|---|---|
| To 9 Buttons and 1/4 yard Shalloon for Vest @2/10 | 0.4.1 |
| To Maiking you a blew Cloath Coate | 0.16.0 |
| To Silk and thread Buckrom and Hair | 0.5.3 |
| To Mohair and tape | 0.1.0 |
| To 4 yards of London Shalloon for Coate Lyning @e/4 | 0.13.4 |
| To 16 Buttons | 0.2.0 |
| To 1 yard of Lynnon for Sleaves and pockits | 0.2.7 |

April 2, 1767

| | |
|---|---|
| To Mending your Brown Cloath Coate | 0.3.6 |
| To Sleave Lynings for Coate | 0.1.6 |

June 9

| | |
|---|---|
| To maiking you a pershon Vest | 0.8.0 |
| To Silk and thread Buckrom and Hair | 0.3.6 |
| To 12 Buttons | 0.0.9 |
| To 1 yard of Lynnon for Vest Body and pockit | 0.2.8 |
| To Maiking your Son Benjamin Together Coat Blew Cloth Brichis. | 0.16.0 |
| To 3 yards of Durant For Coate Lyning. @2.8 | 0.8.0 |
| To ½ ounce Silk for Coate and Blew Cloth Brichis | 0.2.6 |
| To thread Buckrom and Hair | 0.2.6 |

| | |
|---|---|
| To Mohair and tape | 0.1.6 |
| To 14 Coate Buttons | 0.1.9 |
| To 16 Vest DO for Brichis | 0.1.0 |
| To Lynings for His Brichis | 0.2.10 |

26

| | |
|---|---|
| To 1 ¼ yards Lynnon for Coate | 0.3.4 |
| Body Sleaves and Pockits | 2/8 |
| To Maiking you a pair of Black Nit Brichis | 0.6.0 |
| To Silk and thread Buckrom and Hair | 0.2.6 |
| To Mohair and tape and faicings | 0.1.9 |
| To Lynings 3/9 and a pair Silk Gartors @2/6 | 0.6.3 |
| To 18 Buttons | 0.1.1 |
| Delivered January of 22, 1771 | £25.18.0 |
| May of 28, 1773 to cash paid Mr Duffel in full on ? | <u>6.19.0</u> |
| | £32.17.0 |

CHAPTER SEVEN

# Edward Duffield's Philadelphia

*Philadelphia's growth and prosperity were generated by the shrewd merchants and skilled artisans who lived and worked there.... And both (Penn and Franklin) promoted the religious freedom, economic opportunity, and participatory politics that made this eighteenth-century colony, in Penn's prophetic words, "The seed of a nation."*

— Richard S. Dunn, "Religion, Politics, and Economics, Pennsylvania in the Atlantic World, 1680–1755," *Worldly Goods*, Philadelphia, 1999, p.32.

Edward Duffield's life and career were molded by his exceptional city. Philadelphia was founded later than Boston, New York, and Charleston, and it was laid out, planned, and governed based upon progressive and tolerant ideals. By the years of Edward's adulthood it had grown to be the most prosperous and largest of North American cities. It played pivotal roles during the lead-up to our Revolution, our long battle for independence from England, and the first years of our new republic. During the final decades of Edward's life, Philadelphia was also the nation's center for science and culture as well as politics. This offered him many opportunities to observe and participate in important advances and events. No biography of Edward Duffield could be comprehensive without these understandings.

As a prominent wealthy citizen deeply involved in civic and church affairs, as well as in his horological trade, Edward likely was personally involved in historic happenings occurring a few steps from his door. His and his family's long association with Benjamin Franklin, and with Franklin's wife and daughter, drew Edward even closer to momentous occasions. For example, the first meeting of Franklin with Thomas Jefferson and the Committee of Five, responsible for drafting the Declaration of Independence, happened at Edward's Benfield estate.

The city's story began in 1681, about fifty years before Edward's birth, when King Charles II granted fervent Quaker William Penn a large portion of the monarch's North American lands to repay massive debts owed to Penn's father, Admiral William Penn. The territory included much of present-day Pennsylvania and Delaware. Penn sold deeds of settlement to prospective landowners and colonists. One of them was famed London Quaker clockmaker Daniel Quare, who purchased a deed but, unfortunately for American horology, never emigrated. He was one of the "first purchasers" in May 1682, buying 250 acres. In 1699, he paid Penn £2,000, along with three partners, for leases of more land owned by their joint-stock Pennsylvania Land Company of London, which operated until 1760.[261]

*Detail from replica of a 1750 map of Philadelphia by N. Scull and G. Heap. Author's collection.*

Penn himself travelled to his colony, arriving in 1682, but staying for just twenty-two months. He signed a legendary accord with Lenni-Lenape people under the Treaty Elm ensuring peace for his colony. He interested himself in the layout of the city and established it according to a grid of wide straight streets, large lots, and brick buildings. This lowered the risks of fires and epidemics afflicting London, Boston, and other cities that had crowded haphazard layouts and wood-frame construction. Penn envisioned "a greene Country Towne, which will never be burnt, and always be wholsome."[262] He established basic public education and bounties for immigrant artisans, including those who made sundials, and mandated that, "the hours for work and Meals to Labourers are fixt, and known by Ring of Bell."[263] Bell-ringing times often were determined by sundials and hourglasses, and by 1700, a bell-ringer was ordered, "to go round ye town with a small bell in night time, to give notice of ye time of night & the weather, & if any disorders or danger happen by fire or otherwise in the night time, to acquaint the Constables yrof."[264] At night, when sundials were useless, the ringer may have been issued a reliable pocket watch.

Bell-ringing times were a necessity in a colony where many inhabitants did not have access to mechanical timekeepers and instead made use of sundials, hourglasses and almanacs. As late as 1786, a writer claimed that almanacs were still being consulted by:

> [N]ine-tenths of mankind and in fair weather were more sure and regular than the best timepiece.... Twenty gentlemen in company will hardly be able, by the help of their 30-guinea watches, to guess within two hours the true time of night, whilst, the poor peasant, who never saw a watch will tell the time to a fraction by the rising and setting of the moon and some of the particular stars, which he learns from his almanac.[265]

Hourglasses, which cost much less than watches, did not tell time but merely timed intervals. Merchant John Redman's counting house in 1722 had 284 hour and half-hour glasses.[266] In 1730, Benjamin Paschall's shop had a dozen half-hour sandglasses, and merchant Samuel Boude had three dozen.[267] Thirty-minute sandglasses were also commonly used at sea to time shipboard "watches" and on land to limit church sermons unless the minister was moved to "take another glass."

That summer of 1730, twenty-one ships carrying 2,000 immigrants from England arrived. Sixty more ships docked in 1683 bringing 2,000 more people. Another 1,000 landed during a six-week period near the end of the same year, straining supplies and infrastructure. Two years later, 150 enslaved Africans arrived in a ship commissioned by an English mercantile house. They were promptly sold. By 1685 the wave of immigration ended after fifty ships delivered another 4,000 settlers. Not all Quakers, they were recruited in England to populate the colony and they included indentured servants whose Atlantic passage was repaid with several years of unpaid labor. Many arrived impoverished and sick, or soon became so, and they relied on Quaker commitments to caring for the less fortunate.

Gabriel Thomas returned to England after fifteen years in the colony, having arrived in 1681. His published account, based purely on personal experiences, described Pennsylvania's beauty, prosperity, and bountiful resources. Watchmakers and clockmakers were included in his lists of flourishing trades. He encouraged England's poor people to emigrate and added: "Of Lawyers and Physicians I shall say nothing, because this Countrey is very Peaceable and Healthy; long may it so continue and never have occasion for the Tongue of the one, nor the Pen of the other, both equally destructive to Mens Estates and Lives; besides forsooth, they, Hang Man like, have a License to Murder and make Mischief."[268]

In the next forty or so years, Philadelphia grew, with numerous public buildings being added to the cityscape. In 1707 the Court House or Provincial Hall, two blocks from Edward's future home, became the hub of judicial, civic, and mercantile activity:

**Figure 7.1**  View of Second Street, North From Market Street, About 1800. W, Birch & Son. Courtesy of the Library Company of Philadelphia

It stood in the middle of Market street at the corner of Second and back of it the market sheds or shambles stretch away towards the west, occupying the whole middle of the street, and increasing in extent year by year as the city grew and more accommodations for the farmers became necessary. It was a substantial brick structure, built on arches, and was similar in character and appearance to the town halls of that day in many English country towns…. Monarchs on their accessions were there proclaimed; wars were thence declared; and peace, when it came to bless the people, there found a voice to utter it. New governours addressed the people over whom they were appointed to rule, from its balcony… There centered all the official, legislative and administrative life of the Province, there the Provincial Council sat, there the elections were held and there were the goal and those much dreaded but effective instruments of correction, the pillory, the stocks and the whipping post.[269]

The building can be seen prominently on the left of Birch's view of Second Street (Figure 7.1) and on page 77. Edward's corner was slightly visible further down the wide thoroughfare where "passed all the traffic for New York and every place to the north."[270]

In October 1735, the Assembly occupied the newly-completed State House (now Independence Hall) where Edward later maintained the public clock. An early print illustrated the building as Edward would have known it (Figure 7.2).

In 1740, at Fourth and Arch streets, followers of preacher George Whitefield built the first

**Figure 7.2** J. Rogers print of Pennsylvania State House. Author's collection.

tabernacle in this country for a religious evangelist. The New Building was a showpiece and the scene of public gatherings. Edward's Anglican family, however, was firmly attached to Christ Church that was attended by much of the city's non-Quaker elite. With its 200-foot tower, it was one of the largest and most attractive church buildings in North America. Edward lived a short walk away.

In 1744, when Edward was fourteen years old, a Scottish physician, Alexander Hamilton, made a 1,624-mile summer excursion in North America, which included several days in Philadelphia. His book provided details about Edward's city during a hot summer and gives us a sense of what it must have been like for Edward growing up there.[271] He complained that shops below his boarding-house room were noisily opened as early as five a.m. and that most people were staid, devoted to business, and uninterested in luxury or imbibing. Contrary to his expectations, Hamilton found most city buildings "low and mean,"[272] and the streets dirty, unpaved, and obstructed with trash and debris. He saw painted awnings that provided shade over windows and doors, and balconies where gentlemen sat and smoked. Women rarely appeared in public except for church and public meetings. He mentioned the city's public clock that struck the hours on a bell at the bustling public market. He suggested that the city's simplicity was due to Quaker aversion to showy displays.

By 1730, when Edward was born, Philadelphia counted 12,240 inhabitants and was densely urban with residents crowded into neighborhoods near the river and wharves. Indentured servants were increasingly a fixture of Philadelphia life and added to these numbers. During 1732 and 1733, eighteen ships arrived, mainly with Germans and Scots-Irish. Ship captains auctioned those passengers individually to local residents or in groups to merchants for resale. Tens of thousands more arrived during the next two decades. Some remained in the city while most headed west to rural and frontier regions.

Because of the influx of new inhabitants to the city, Penn's large lots with extended rear gardens became subdivided into alleys and rows of smaller houses in which tradesmen lived above their first-floor shops. A 1705 ordinance, similar to ones in London, had prohibited men not admitted as freemen from keeping open shops or practicing their trades. This restriction was ignored and forgotten, with no effect on the city's artisans, whose numbers had grown rapidly since the original ordinance.

Crowded conditions led to epidemics, which Penn had hoped to avoid through his careful city planning. Smallpox often ran rampant: Franklin wrote "The Small-pox has now quite left this City. The Number of those that died here of the Distemper, is exactly 288, and no more. 64 of the Number were Negroes; If these may be valued one with another at £30 per Head, the Loss to the City in that Article is nearly £2000."[273] It may have been one of these epidemics that carried off Edward's mother and four siblings.

In the 1740s, Edward's teenage years, Philadelphia grew further and was also touched by events in Europe. On June 11, 1744, the War of the Austrian Succession pitted England and its colonies against France. It was formally proclaimed at the Philadelphia courthouse and citizens were urged to arm themselves. The announcement triggered a procession of more than 4,000 enthusiastic residents cheered by ladies and gentlemen from windows and balconies. Fourteen-year-old Edward may have marched or observed.

A year later, the supposedly impregnable French fortress of Louisbourg in Nova Scotia was captured on June 28, 1745, by a force of New England militiamen and British naval units. This unlikely victory was celebrated in Philadelphia with rejoicing, parades, bonfires, and multiple toasts throughout the city. Colonial Governor George Thomas requested the pacifist Quaker government to send supplies to the British forces. The Assembly agreed only to supply foodstuffs, including "other grain," which the Governor craftily construed as gunpowder. Fearful of French privateer attacks, Franklin and supporters organized a lottery to fund a defensive artillery battery. The pacifist Quaker government again refused military expenditures, leading Franklin to publish the pamphlet *Plain Truth* in support of defending the city. Hundreds of Associators formed into militia companies. Edward was too young to enlist.

Peter Kalm, a Swedish botanist, arrived in Philadelphia on September 15, 1748. He, too, described its thoroughfares, and his description gives us a sense of how Philadelphia looked at the time:

> The streets are regular, pretty, and most of them fifty feet.... Some are paved, others are not, and it seems less necessary since the ground is sandy and therefore soon absorbs the wet. But in most of the streets is a pavement of flags, a fathom or more broad, laid before the houses, and four-foot posts put on the outside three or four fathoms apart. Those who walk on foot use the flat stones, but riders and teams use the middle of the street. The above-mentioned posts prevent horses and wagons from injuring the pedestrians inside the posts, and are thereby secure from careless teamsters and the dirt which is thrown up by horses and carts....[274]

The botanist Kalm wrote of woods used by cabinetmakers, chiefly black walnut, wild cherry, and curled maple. These woods would also have been used for cases for the clocks that Edward would soon begin making, although Caribbean mahogany became popular for furniture and clock cases later when it was more available and less costly.

Other reports, however, still complained of dangerous streets. There were deep potholes, some caused by underground creeks, and stumps and rank trash heaps that obstructed passage. During the spring and summer of 1746, there was an epidemic of "putrid throat"[275] (probably diptheria) that teenage Edward may have avoided by staying at Benfield. By 1749, the city and its nearby suburbs recorded 2,076 houses and its busy port cleared approximately 400 vessels a year. German immigration reached its peak during the next five years, with more than one hundred ships bringing thousands of new residents.

Cultural opportunities also were expanding during these years. The Union Library Company was formed in 1747 by tradesmen and craftsmen as an alternative to Franklin's Library Company that raised its share price from forty shillings to £9. This second subscription library could have offered Edward additional educational materials. A third subscription library was formed by the Quakers, the Association Library of Philadelphia that charged just twenty shillings per year. Alongside the libraries, educational opportunities also grew. Educator Andrew Lamb in 1748 advertised his many areas of instruction including "surveying gauging, dialling, and astronomy, in great variety.... at the corner house of Chestnut and Second Streets, over against Mr. Stretch's, Clockmaker."[276] Ten years later he still was teaching, then at "John Wood's House, Clock-maker."[277] A new form of entertainment also became available, when dancing assemblies began in the city for the upper classes in 1749. Partners were selected by lot and stayed together for the evening. Teenage Edward possibly danced with his future wife, Catherine, and other young women, at these assemblies.

Philadelphia's population of 25,000 in 1750 made it the largest colonial city in North America. Gottlieb Mittelberger, author of *Journey to Pennsylvania*, arrived on September 29 as one of nearly 500 passengers on the *Osgood*. He took the required oath of allegiance and wrote in detail about life in the colony:

> In Pennsylvania no profession or craft needs to constitute itself into a guild. Everyone may engage in any commercial or speculative ventures, according to choice and ability. And if someone wishes or is able to carry on ten occupations at one and the same time, then nobody is allowed to prevent it. And if, for example, a lad learns his skill or craft as an apprentice or even on his own, he can then pass for a master and may marry whenever he chooses. It is an admirable thing that young people born in this new county are easily taught, clever, and skillful. For many of them have only to look at and examine a work of skill or art a few times before being able to imitate it perfectly. Whereas in Germany it would take most people several years of study to do the same. But in America many have the ability to produce even the most elaborate objects in a short span of time. When these young people have attended school for half a year, they are generally able to read anything...[278]

> Many kinds of beverages are available in Pennsylvania and the other English colonies. First of all, excellent and salubrious water; secondly, people drink a mixture of three parts of water and one part of milk; thirdly, there is a good cider; fourthly, small beer; fifth, delicious English beer, strong and sweet; sixth, so called punch, made of three parts water and one part West Indian rum (where there is no rum, one can use brandy, but rum is much more pleasant), mixed in with sugar and lemon juice; seventh, sangaree, which is even more delicious to drink - this is made out of two parts of water and one of Spanish wine with sugar and nutmeg; and then, eighth, German and Spanish wines, which are obtained in all taverns....[279]

> All houses have two benches on each side, set up about four feet straight out in front of the doors. Resting on two columns over each bench is a roof like that of a garden pavilion.... Every evening when the weather is fine people sit on the benches, or promenade in front of them. The streets and houses of this city are so straight that one can look directly ahead for the distance of a half hour's walk.[280]

Gottlieb noted independence and freethinking—"Pennsylvania is heaven for farmers, paradise for artisans, and hell for officials and preachers"[281]—and he observed that "almost everyone, farmers as well as private persons, makes use of silver watches: they are very generally worn by the English

**Figure 7.3** 1784 George Inn, David J. Kennedy watercolor. Collection of the Historical Society of Pennsylvania, Call Number K:5-73.

ladies."[282] Watch ownership was becoming more common and provided expanding markets for imports, sales, and service.

In the mid-eighteenth century, Philadelphia added new educational institutions and a hospital, all of which would have contributed to its citizens' quality of life. In 1751, with Franklin's strong encouragement, the Academy of Philadelphia, later the University of Pennsylvania, opened in the building constructed a decade earlier for Reverend Whitefield. Prominent builder Robert Smith, later to work with Edward on the new jail, did the renovations. Classes first were held in William Allen's warehouse at Second and Arch streets, the same crossroads where Edward opened his shop in 1751, diagonally opposite the George Inn on the northeast corner of the busy crossroads (Figure 7.3). The hostelry bustled with departing and arriving stagecoaches, providing Edward with plenty of visibility and traffic and perhaps also distracting the school's students from their studies.

As the Academy of Philadelphia was preparing to open, a long list of classical and technical courses including "mathematical Sciences" was announced in December of 1750.[283] In November 1752, when Edward was established on his Arch Street corner, John Clare advertised his instruction on those subjects, including surveying, on Front Street.[284] In 1765, Richard Harrison had opened on Front Street a school "At the Sign of the Mariner" to teach "… Surveying, Dialling (sundial-making)… and correcting a Time-piece at any Time between the Hours of 8 in the forenoon, and 4 in the Afternoon."[285] These all were ways that Edward could have learned science, mathematics, and specifics about horology and surveying.

In 1751, the charter for a public hospital, where Edward's son, Benjamin, later served as a physician, was signed. Patients were treated in a rented home at Market and Fifth streets before the grand hospital buildings were completed. Franklin, a key proponent, cited Biblical admonitions and asserted that medical care was ten times more costly in private homes than in a hospital. In May 1754, Franklin justified its ongoing support and expansion by penning a history of the public hospital.[286] He created the hospital seal, which quoted Scripture: "Take care of him; and whatsoever thou spendest more, when I come again, I will repay thee." The next year on April

**Figure 7.4** Pennsylvania Hospital, 1799, W. Birch & Son. Courtesy of the Library Company of Philadelphia

28 a public ceremony accompanied the laying of the hospital's cornerstone. Edward made a £10 donation[287] and he may have joined the crowd marching from the temporary hospital site. A 1799 view depicts the hospital with its distinctive architecture and completed wings (Figure 7.4).

Philadelphia of this period was a lively town with many social, cultural and "philosophical"—as Edward would have called them—opportunities available to its citizens. In 1752, more than 200 ladies and gentlemen, perhaps including Edward and his new wife, Catherine, turned out for a grand ball at the State House to celebrate the King's birthday. Governor James Hamilton (1710–1783) hosted a supper afterwards in the Long Gallery upstairs, and he boasted that he knew the name of every person in the city.[288] Later on he certainly knew Edward, who repaired Hamilton's watch in 1761 and 1764.

Also in 1752, a transit of Mercury across the face of the sun took place but cloudy weather prevented its viewing. Franklin issued a pamphlet explaining how best to observe the rare celestial event. He advised that timing was critical: "As to the Clock, it should be well regulated to the Sun, by corresponding Altitudes of the Sun, taken many Days before and after, if it be possible, to the End that the true Time may be assigned with the more Certainty and Precision of every Observation, and principally at the Moments of the Emersion."[289]

In *Poor Richard's Almanack* Franklin included an illustration and noted that Mercury "will appear like a small black Patch in a Lady's Face."[290] Seventeen years later in 1769, Edward produced an accurate timepiece for the next Mercury transit (Catalogue No.3).

This was only the first of several important astronomical events that took place in Philadelphia during these decades. A long-awaited transit of Venus on June 6, 1761, was the first of two during the century. It was a watershed event in the history of astronomy since solar-system distances could be calculated, as already was understood. High-accuracy timekeepers and time notations were key to the exacting calculations but were not sufficiently precise that year. Worldwide preparations for better observations quickly began for a second transit eight years later.

Then, on November 15, 1763, Charles Mason and Jeremiah Dixon arrived from England to begin their surveys beginning with establishing Philadelphia's precise global coordinates. Their observatory preceded David Rittenhouse's in Norriton by five years. They were back in Philadelphia in 1768, after completing their "Line," to determine the difference in gravity between there and the Royal Observatory in Greenwich. Not sourcing a locally-made clock, they used a precise British regulator by John Shelton that survived a shipwreck on the voyage from London.[291] A year later, Edward certainly saw Halley's Comet, visible even during daylight hours, when it made its first predicted return to Earth's skies since 1682.

In 1753, on March 10, a bronze bell ordered from England's Whitechapel Foundry was hung in the State House steeple. It immediately cracked from a blow of the clapper. Casting flaws or too-brittle metal were blamed. The local foundry of Pass and Stow melted and recast the so-called Liberty Bell, adding more copper, but its tone was bad. A second recasting was still unsatisfactory, but the bell remained in the steeple, sounded by the pull of a rope. A replacement sent from England was hung instead in a cupola where the hours were struck by Thomas Stretch's new clock movement. Both bells were tuned to E-flat.

Bell chimes for the Christ Church steeple were imported at the same time as the larger bell. These rang on special occasions and as the "butter bells" on Tuesday and Friday evenings to announce the next day's public markets, which were renowned for their fresh provisions.[292/293] Edward, of course, heard all these bells; his home was near the church and he could see its 200-foot steeple from his corner.

Some neighbors did not enjoy the frequent bell-ringing. In fact, in September of 1772, they petitioned that: "They are much incommoded and distressed by the too frequent ringing of the great bell in the steeple of the State-House, the inconvenience of which has been often felt severely when some of the Petitioners Families have been afflicted with Sickness, at which times, from its uncommon size and unusual sound, it is extremely dangerous, and may prove fatal."[294]

A year later, in 1754, at the busy corner of Front and High Streets, printer William Bradford opened the London Coffee House, a meeting place for businessmen, ship captains, and politicians. Edward could have perused its racks of English and American periodicals.

In 1755, Edward may have witnessed a public procession, celebration, and banquet held by Philadelphia's Freemasons to inaugurate their new lodge on Second Street. They heard an oration at Christ Church by William Smith, a Scottish clergyman whom Franklin recruited from New York to lead the new Academy. While Franklin, his close friend, was an important member, Edward's name is not on any Masonic rosters.

In 1756 Franklin spurred formation of the Proprietary Militia or Associated Regiment of Foot of the City of Philadelphia. On March 18, more than one thousand formed into ranks on Society Hill, were reviewed by Colonel Franklin, and paraded down Second Street. The previous day more than six hundred marched and displayed their street-firing abilities. Edward may have observed this display as an onlooker, but he was not listed as a militia member.

*The American Magazine* was established in 1758 by Edward's young acquaintances including Francis Hopkinson, William Plumstead, and Thomas Godfrey. An alternative to English magazines, it was published for just one year. The July issue had a long article and a letter from James Logan about the invention of Godfrey's quadrant and its relation to Hadley's version of the same navigational instrument. In August, Logan claimed that Godfrey, not Hadley, was this instrument's inventor, and Godfrey asked Logan to transmit the design to London in hopes of an award. Logan, however, did nothing until, too late, he read Hadley's description of the same invention in the Royal Society's *Philosophical Transactions*.

Godfrey was a member of Franklin's Leather-Apron Club, or Junto, and calculated ephemerides for Franklin's almanac. His second son, also named Thomas, wrote the *The Prince of Parthia*, the first play by an American author performed upon a regular stage. It was presented in 1759 at the new Southwark Theater. More than 3,000 residents and clergymen petitioned against the theater, including Edward's Presbyterian cousin Reverend George Duffield. Many other influential citizens signed a counter-petition, but Edward signed neither. The senior Godfrey died in 1749, when Edward was still a teenager living with his family, and the newspaper announcement reported that Godfrey "had an uncommon Genius for all kinds of Mathematical Learning, with which he was extremely well acquainted."[295]

Events outside of Philadelphia in these years also would have had an effect on Edward and his fellow citizens. In 1752, England and its dominions adopted the Gregorian calendar, eliminating eleven days when the date jumped from September 3 to 14. Each new year then began on January 1, not as before on March 25. Continental Europe had made the shift several decades earlier, in 1582, to correct for the loss of eleven minutes per year in the outdated Julian calendar. The shift was delayed and resisted in England in part because it had been initiated by Pope Gregory. Anti-Catholic Royal Society member Isaac Newton shelved the matter for many years but eventually his and others' objections were overruled. In Philadelphia, as elsewhere, those skipped dates raised objections and complications regarding pay, debt, and other financial concerns. Edward's tenants certainly were affected, having eleven fewer days to save for the next month's rent. Religious groups complained about missing a Sunday's observance and instruction. *Poor Richard's Almanac* explained the issue and alerted readers to adjust dates from past years.[296]

With the French defeat at Montreal and the war's end in 1760, a severe economic depression took root. Forty thousand British soldiers and seamen, spending their hard-currency wages, were withdrawn. With peaceful seas reopening, more immigrant ships began arriving at the sixty-six wharves that lined Philadelphia's two-mile waterfront. The following year, on September 22, King

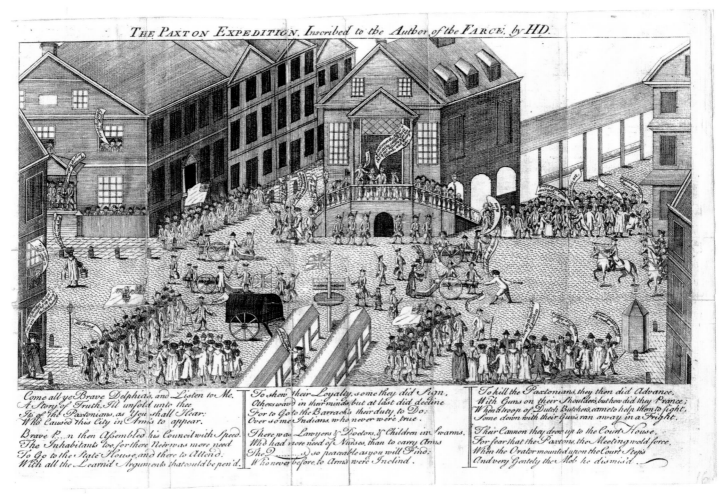

**Figure 7.5** Dawkins, Henry. *The Paxton Expedition: Inscribed to the Author of the Farce*, Philadelphia, 1764. Courtesy of the Library Company of Philadelphia.

George III was crowned and the news was proclaimed to a large Philadelphia crowd that Edward no doubt joined.

The Paxton Boys crisis erupted in 1764. Militiamen mustered to defend against a mob of frontier residents irate about inadequate protection against Indian raids. Franklin intercepted and placated them at Germantown, accompanied by Christ Church Reverend Sturgeon and several others. Edward may have been with them, and he could easily have observed the militiamen nervously gathered in the heart of the city. David Rittenhouse was furious at the Paxton Boys who passed his home:

> About fifty of the scoundrels marched by my work-shop — I have seen hundreds of Indians traveling the country and can with truth affirm, that the behavior of these fellows was ten times more savage and brutal than theirs. Frightening women, by running the muzzles of their guns through windows, swearing and hallooing; attacking men without the least provocation; dragging them by the hair to the ground, and pretending to scalp them; shooting a number of dogs and fowls — these are some of their exploits.[297]

One of the earliest published views of Philadelphia was a satirical print by Henry Dawkins about this crisis (Figure 7.5). He depicted the crowded scene at the Court House.

On March 22, 1765, the British Parliament passed the infamous Stamp Act. The *Pennsylvania Gazette* published the text on April 18, sparking angry opposition and non-importation resolutions. Citizens were strongly urged to patronize local artisans rather than purchase English-made goods. Businesses were issued pre-printed forms to countermand outstanding orders from English vendors, and merchants sent memorials to their English counterparts urging repeal. Edward and other horologists, however, were not on a list of thirty-four artisans vowing non-importation, perhaps because they relied on English supplies. The next year on March 18, the Act was repealed. News reached Philadelphia on May 20. That night the city was illuminated by bonfires. Barrels of beer were rolled out. Toasts were pronounced at a State House banquet for 300 people. Celebrations continued on June 4, the King's birthday, with picnics, salutes fired from wheeled barges, fireworks, and an outdoor evening feast for 430 revelers.

As the 1770s began, events continued to make an impression on Philadelphia's residents. Near Edward's home the residence of cabinetmaker William Savery, perhaps a maker of cases for Edward's clocks, was badly damaged by a 1772 fire that began in an adjacent shop. Edward may have been among the locals praised in a newspaper account for battling the fire during severe hail and snow.[298] By local subscription in 1773, the City Tavern was erected on Second Street between Chestnut and Walnut, supervised by a Board and managed by Daniel Smith. Edward could have strolled there to drink coffee, converse with acquaintances, and read newspapers and magazines. Dr. Benjamin Rush delivered the APS's annual oration in 1774. This was the city's main intellectual event and almost certainly Edward, an APS member, was present. The next February, David Rittenhouse was selected to be the orator. He spoke about the history of astronomy and revealed his broader beliefs about nature, the rights of man, liberty and self-government, and the wrongs of slavery.[299] The APS ordered the speech printed and distributed to delegates of the Second Continental Congress.

At the same time, Philadelphia, like the rest of the American colonies, became increasingly embroiled in the events that led to the Revolution. Already in 1770, debates intensified over whether to continue non-importation. Rightly predicting a flood of cheap English goods, artisans organized to maintain the policy. Merchants argued that non-importation hurt the local economy as other colonies resumed trading and British counterparts provided bullion and credit. As both merchant and artisan, Edward could have been on either side of the issue.

December 13, 2023, was the 250th anniversary of the 1773 Boston Tea Party. On June 1, 1774, Philadelphia residents expressed outrage against the British closing of Boston's port in retaliation for the event. Shops were closed. Christ Church's muffled bells pealed solemnly from morning to night. Twelve hundred mechanics met on June 9 to hear a letter from New York. Another mass meeting on June 10 placed Edward on a committee of local mechanics, along with David Rittenhouse, to develop resolves for another mass meeting eight days later.[300] Paul Revere arrived from Boston and perhaps was introduced to Edward.

At five in the afternoon of April 24, 1775, an express rider, Israel Bissell (1752–1823) completed a marathon ride from Watertown, Massachusetts. He galloped into Philadelphia bringing news of the battle of Lexington and Concord. He carried a handwritten message that was copied and printed at each stop:

> Wednesday morning near 10 of the clock—Watertown. To all the friends of American liberty be it known that this morning before break of day, a brigade, consisting of about 1,000 to 1,200 men landed at Phip's Farm at Cambridge and marched to Lexington, where they found a company of our colony militia in arms, upon whom they fired without any provocation and killed six men and wounded four others. By an express from Boston, we find another brigade are now upon their march from Boston supposed to be about 1,000. The Bearer, Israel Bissell, is charged to alarm the country quite to Connecticut and all persons are desired to furnish him with fresh horses as they may be needed. I have spoken with several persons who have seen the dead and wounded. Pray let the delegates from this colony to Connecticut see this.[301]

The next morning, Edward would have been among the 8,000 residents who swiftly assembled at the State House. They approved a strong resolution and began forming local militias. In June,

Congress appointed George Washington to lead its troops, and thousands of city residents gathered to cheer his departure to take command of the siege of Boston.

The Declaration of Independence was first discussed in June by the Committee of Five who met at Edward's Benfield estate, where Franklin was staying to recover from a severe gout attack. The final version was proclaimed by Sheriff John Nixon. Philadelphia merchant Charles Biddle wrote disdainfully: "I was in the State House Yard when the Declaration of Independence was read. There were very few respectable people present."[302] John Adams reported that Christ Church chimes were pealing despite its Anglican members' questionable support for the Revolution.

The city's Committee of Safety—David Rittenhouse was its Vice-President—sent teams of men throughout the area to seize lead to mold into musket balls and bullets. Clock and window weights were primary sources of the soft heavy metal. On one foray, the men discovered that the window weights at the Drinker residence were iron, not lead, and were left behind. A New Jersey clockmaker avoided seizures when his Township of Hillsborough decided to "leave in the hands of Mr. Isaac Brokaw, clockmaker, thirty pounds weight of lead, he having represented to the convention that he could not carry on his trade without such quantity."[303]

On July 20, a general fast day was declared by Congress. Edward likely was in his pew at Christ Church to hear a sermon by Thomas Coombe. This cleric preached restraint and later was imprisoned for Loyalist views.

On September 26, 1777, 3,000 British troops led by General Charles Cornwallis occupied Philadelphia, marching down Second Street and camping on Society Hill. One of the final orders of Congress before fleeing was to remove the city's bronze bells to prevent the British from melting them. Reverend Jacob Duché, chaplain to the Continental Congress, protested that taking them down involved great risk, but the bells were safely hustled away to Allentown.

During the occupation, many of the rebel merchants' shops were taken over by Loyalists. British Major John André occupied Franklin's Market Street home, where his daughter Sarah and her family had been living. Before fleeing, she removed books and valuables but not mahogany furniture. When they returned the following summer, they prepared a list of many missing and damaged items. Thomas Paine reported in a letter to Franklin that much of that furniture, and perhaps clocks, had been sold or destroyed. During the occupation, Sarah's family spent time with the Duffields, north of the city's military perimeter.

The British withdrew next June with nothing accomplished by holding the Revolution's capital. Three thousand fearful Loyalists joined the exodus. The city was in derelict condition with many properties damaged and looted. Edward was not listed among owners, including cabinetmaker Benjamin Randolph, who reported more than £1,000 in losses. Either Edward's properties were spared, or he chose not to report on damage to them.

After this, life in Philadelphia remained relatively untouched by the remainder of the war, but other events, both local and global, affected Edward. In 1780, the Gradual Abolition Act passed in Pennsylvania as the first anti-slavery law in the new United States. It stated that children born to enslaved mothers were free, but were required to work unpaid until reaching age twenty-eight. Martindale writes that Edward's deceased enslaved workers were "buried at the end of the lane near some lilac bushes."[304] As a slaveowner, Edward was affected by this law.

Then in 1784 the eruption of Iceland's Laki volcano caused crop failures across Europe and North America and a drop in global temperatures. The "Laki Haze" was one of the most damaging climatic events of the millennium. North America's winter that year was one of the coldest on record. It brought the longest period of below-zero temperatures in New England, the largest accumulation of snow in New Jersey, and the lengthiest freezing of Chesapeake Bay. A huge snowstorm hit the South, the Mississippi River froze at New Orleans, and there were ice floes in the Gulf of Mexico. This climate disaster certainly impacted Edward's livelihood. The economy slid into recession, European imports surged, cash and hard currency disappeared, and commerce radically declined. A forced sale of the goods and properties of Edward's tenant Luke Keating took place in November. In a 1784 lecture, Benjamin Franklin recounted:

> During several of the summer months of the year 1783, when the effect of the sun's rays to heat the earth in these northern regions should have been greater, there existed a constant fog over all Europe, and a great part of North America. This fog was of a permanent nature; it was dry, and the rays of the sun seemed to have little effect towards dissipating it, as they easily do a moist fog, arising from water. They were indeed rendered so faint in passing through it, that when collected in the focus of a burning glass they would scarce kindle brown paper. Of course, their summer effect in heating the Earth was exceedingly diminished. Hence the surface was early frozen. Hence the first snows remained on it unmelted and received continual additions. Hence the air was more chilled, and the winds more severely cold. Hence perhaps the winter of 1783–4 was more severe than any that had happened for many years.[305]

Despite that year's hardships, there were still social and cultural opportunities to be had. In July, thousands gathered in Potter's Field, later Washington Square, to observe a balloon ascension from the yard of the new jail. Edward's son, Benjamin, was among the venture's subscribers and Edward may have accompanied him. The silk balloon, thirty-five feet in diameter, had a furnace below to generate hot air. As it ascended, the basket knocked against the prison wall, ejected the balloonist, and rose rapidly into the sky without him. It suddenly burst into flames and the crowd wrongly assumed that the aeronaut still was on board and was incinerated.

In 1788, on July 4, the Grand Federal Procession celebrated the new United States Constitution. Christ Church bells pealed to greet the dawn, cannons roared from the ship *Rising Sun*. Five thousand spectators lined the one-and-a-half-mile parade route to applaud marchers representing eighty-eight city societies. Twenty-four clock- and watchmakers constituted the thirty-ninth group, but there is no record that Edward was among them. They were headed by John Wood, Jr. A silk flag displayed the arms of the London Clockmakers' Company, and its Latin motto translated as "Time Rules All Things."[306] The Agricultural Society was headed by its president, Samuel Powell. Cabinet and chair makers marched, led by Jonathan Gostelowe who held dividers, and was followed by a rolling workshop manned by apprentices and journeymen. Committee chairman Francis Hopkinson published a detailed report that described the entire event.

Edward would have joined seventeen thousand others that evening at a picnic on the Bush Hill mansion grounds. No spirituous liquors were allowed, but revelers quaffed porter, beer, and cider to wash down a cold collation served on a huge circle of tables. At each toast, a trumpet sounded and was answered by ten blasts of artillery.

In January of 1793, a crowd at the jail observed another hot-air balloon accession. Paying guests included President Washington and the four Founding Fathers who succeeded him: John Adams, Thomas Jefferson, James Madison, and James Monroe. Jacob Hiltzheimer was a witness, as perhaps was Edward. Forty-thousand spectators watched for free from outside the walls. French aeronaut Jean-Pierre Blanchard carried a signed letter from Washington to whomever might receive it, the first letter sent by airmail.

In August of that year, disaster struck with an epidemic of yellow fever. Exacerbated by a lengthy drought, the disease killed more than 4,000 city residents. One was John Wood, Jr. who died October 9. His gravestone in Old Saint Paul's Episcopal Church Burial Ground was inscribed "Clock & Watchmaker." Three other horologists succumbed—John Dalton, Chas. Fk. Dubois, and John Stillas[307]—as named in printer Mathew Carey's account on the tragedy that listed, in alphabetical order, the city's thousands of deaths.

Carey's published report stated: "The watches and clocks in this city, during the disorder, were almost always wrong. Hardly any of the watchmakers remained and few people paid attention to how time passed. One night, the watchmen cried ten o'clock when it was only nine, and continued the mistake all the succeeding hours."[308]

Living outside the crowded city, Edward was not in as great a danger. We do not know if he harbored on his estate any of the twenty-thousand-plus city residents, mostly the better-off, who fled during the epidemic. Lucy and Erenna Duffield were listed among the victims, but their connection, if any, to Edward is unknown. If, prior to that August, anyone died who was residing with Edward at Benfield, their names would not have been recorded by Carey who listed only deaths of city residents.

Edward's will indicated that at the time of his passing ten years later he still owned his "two dwelling houses" at the corner of Arch and Second streets within the city limits. He may have also owned and rented other city properties at the time of the epidemic. Most businesses in the city came to a standstill or closed, and, except for coffin-making and grave-digging, most laboring and craft jobs disappeared. Carey applauded landlords who waived or delayed rent payments for tenant families unable to pay or whose breadwinners had perished, and the publisher castigated landlords who heartlessly evicted and seized clothing and personal property from indigent tenants. The records do not tell what Edward chose to do with his tenants.

As for the cause of the disease, Dr. Benjamin Rush initially blamed putrid coffee rotting on Ball's Wharf. Edward's son Benjamin questioned this theory and earned a harsh rebuke from his friend Rush. The younger Duffield attempted to make amends, writing to Rush: "If to differ from you in medical Opinions is a Crime, it is one that every man must at one Time or other of his Life, be guilty of with some of his Fellow Travellers in the Difficult Road of Science."[309]

Within Rush's Commonplace Book, 1792–1813, he included Benjamin's name in entries titled "Pupils' ingratitude."[310] A footnote by the editor of the transcriptions claimed, however, that "Duffield was a notably sociable and hospitable man."

There were other bitter public disputes among local physicians about possible remedies. Dr. Rush's argument with Dr. Adam Kuhn and his followers was particularly vitriolic. Rush insisted upon copious bleedings and large purgative doses of mercury—now known to be exceedingly harmful although many of his patients somehow survived—while Kuhn favored treating with bark or chamomile tea, cold baths, wine, and no purging or bleeding. Edward's son agreed with Kuhn and Bush Hill's Dr. Devèze, provoking Rush to write to Redman Coxe: "B. Duffield seldom concludes a lecture without a phillipic against me."[311]

Desperate citizens fumigated rooms, soaked handkerchiefs with vinegar, wore bags of camphor around their necks, burned gunpowder and tobacco, lit bonfires, and fired small cannons in the streets. Benjamin suggested spreading a two-inch layer of fresh earth in a room and renewing it daily. He stated that this would be a comfortable and sure antidote if supplemented by frequent warm baths and "the Asiatic remedy of myrrh and black pepper."[312] We do not know if his father tried any of this.

Many of the city's public officials, politicians, clergymen, merchants, and physicians chose to flee, not knowing that the disease is mosquito-borne and cannot be passed from contact with the sick or dead. Still, separation reduced the ease of transmission and residing outside of Philadelphia was marginally safer. Among the thousands of victims, rich and poor, were Dolley Madison's first husband and the daughter of clockmaker Owen Biddle.

The epidemic faded after November frosts but returned with the next year's warm weather. In its second year, it killed approximately 1,000 inhabitants, fewer than the previous year due to increased immunity and better precautions. The new outbreak was blamed by some residents on the opening of the New Theatre modeled on the Theatre Royal in Bath, England. They claimed that the theater's sinfulness brought divine punishment.

Sickness again afflicted the city in 1795. The first reported victim was U.S. Attorney General William Bradford, Jr. Cabinetmakers Thomas Affleck and Jonathan Gostelowe also died. Two years later, another outbreak caused more deaths and a general exodus. Federal offices moved away from Philadelphia, and President John Adams fled home to Braintree, Massachusetts. By the time the pestilence subsided, 1,292 more had died. Edward's son Benjamin sickened but recovered, unlike nine city doctors who perished.

Jacob Hiltzheimer's final diary entry was for September 5, 1798; he died of yellow fever nine days later. Retired instrument maker, Benjamin Condy, also perished that August after working in the city since the 1750s and no doubt known to Edward. Franklin's grandson Benny Bache, who had not fled the city, succumbed. If he had survived, he would have been the first man tried under the Alien and Sedition Acts for his newspaper's attacks on Federalist officeholders.

Charles Willson Peale retreated with his family and servants to New York City. Nearly all were sickened but recovered, except for the enslaved Titian, who died on September 18. Their Black servant was named Nancy Duffield but her possible connection to Edward's family is not known.

Diarist Elizabeth Drinker fled her home during another outbreak. She wrote that "2 clocks which were wound up when we left home, every hour give the time to the insects and mice if any there be."[313]

On a happier note, Peale in 1802 rented a substantial portion of the State House for his gallery of 200 stuffed animals, 1,000 birds, 4,000 specimens of insects, and collections of minerals, fishes, and snakes. Another room exhibited more than a hundred portraits of famous statesmen and soldiers, painted by himself and his son Rembrandt. Most of those faces would have been quite familiar to Edward, who may have enjoyed touring the gallery during this last year before his death.

From the time of Edward's birth in 1730 until his passing in 1803, momentous events took place in his city. Some were tragic like rampant epidemics, others were earthshaking like the American Revolution and our new nation's Constitution. Edward's front-row seat, his friendship and association with principal movers, and his direct participation in civic and church affairs all amplify and justify this comprehensive examination of his life in one of North America's most important population, financial, and political centers.

Names visible on the map (approximate reading, top to bottom, left to right):

- Matthe[w] C[ ], W[m] Salaway, Samuel Claridg
- [Hum]phry Merry, W[m] Powell, Israel Iobs, Silas Crispin
- [Jo.] Phillips, John Rush, Phil. Tleaman
- Marey [J]efferson, W[m]. Sherter, Eliz. Martin
- William Trampton, Iosh Cart, Samuel Allen, Joel Ielson
- Ino. Russell, John Barnes, Trick Holme, Richard Wood
- Pet. Robinson, Sarah Fuller, Danl. Heaphy, Tho. Fairman
- Richard Will, Richard Wall, Ios Phips
- Tob. Deach, Tobias Leach, W[m] Chamberlin, W[m] Stanly
- John Ashman, Rich Dungworth, Silas Crispin
- Everard Bolton, Walter King
- W[m] Brown, Robert Fairman, John Mason
- John Day, Tho. Levesly
- Nehemiah Michell, Rich Worrell, Robert Fairman
- John Wes, Alin Foster
- Benjamin East, John Simmer, Iam Atkinson
- Henry Waddy, John Bunto, Ios Ashlet, Ios Fisher
- John Hughs, John Tarnes, Kat. Marten, Rob. Turner
- John Harper, Sam. Claridg
- Rob Addams, Tho. Ducket, Tho. Holme
- Henry Wadd, Thomas Sare, Elenor Holme
- Thomas Fairman, Ch[t] Thom[as], Pet. Rambo Ju[r]
- Toaconing Township, Enoch G Salter, An Salter, Lass Borr, Ino Gilbert, Joseph Growden
- Erick Cock, W[m] Salway, Georg Foreman, Knoch Harrison, George Keene, Ch[as] Chas, Erick Meels, Pet. Rambo, Ben Acrud, Walter Forest
- Monas Cock, Neels Nelson
- Giles Knight, John Tibby, Thomas Cross, Samuel Wells, Daniel Jones, Andrew Griscomb, George John, Collis Hart

The Mannor of Moreland

Dublin Cr.

Potquessin C.

DELLAWARE RIV[ER]

CHAPTER EIGHT

# Edward Duffield's Properties and Lands

*TO BE SOLD, A Valuable Plantation, ... about 225 acres of very good land, well improved, with one good dwelling house, suitable for a genteel family; a good barn, stabling, coach house, and a great variety of other useful and convenient buildings; also another dwelling house and out-buildings, at a convenient distance, to accommodate a tenant. There is a good young bearing orchard, and a large proportion of meadow grounds, with a durable stream of water running through the same.*

— Ad placed by Edward Duffield, March 6, 1795

Edward's long life centered on three properties: his grandfather's rural Benfield estate, where he was born and lived during his final decades; his parent's Market Street home next door to Franklin's in-laws; and his own home at Second and Mulberry (later Arch) streets, where he spent twenty-four years as a merchant and horologist.

Many details about those residences are included below, thanks to insurance surveys, tax records, and census data. Included for context and comparisons are similar records for family members, friends and associates, and fellow tradesmen. Not every cited detail of his assessments, taxes, and holdings has been end-noted. These would be voluminous and repetitive, and they can be verified in online and published city, county, and state records. Pre-internet published compilations were consulted and cited. These include the exhaustive published research by Hannah Benner Roach (1907–1976) and the easily-accessed 1769 Philadelphia County and City Proprietary Tax List that was reprinted in 1988 from the Pennsylvania Archives Third Series.

Edward's wealth placed him in society's upper echelons, well above most of his artisan colleagues. He was in the uncommon position of being a "mechanic" among the city's elite, reinforcing his significance as an eighteenth-century gentleman craftsman. This was different from upper-class Englishmen who pursued tinkering and experimentation but never for pay.

Edward in 1751 reached age twenty-one and began his married working life. Almost nothing about him is known before that date except for his birth, inheritances from his grandfather, and the tragic loss of his mother and brothers when he was six years old. When he came of age he received from his father, as stipulated by his late grandfather, the deed to buildings and land measuring forty-two feet by fifty-one feet at the northeast corner of Second and Mulberry streets, where he lived until 1775.

*Fac-simile of a Portion of Holme's Map of the Province of Pennsylvania With Names of Original Purchasers from William Penn. 1681.*

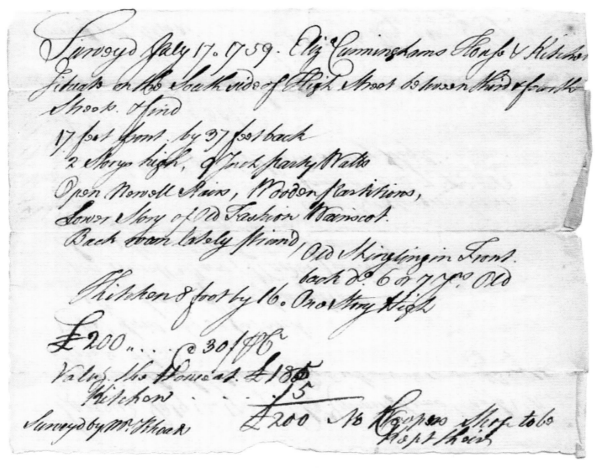

**Figure 8.1** 1759 survey of the former Duffield home next door to Franklin's in-laws, Insurance Survey S00430, Local ID Number: S00430-001A. Courtesy of the Contributionship for the Insurance of Homes from Loss by Fire.

That plot initially was purchased in 1683 by John Farmer from Thomas Holme, an Irish Quaker who William Penn had made surveyor-general to lay out the city's overall plan. Edward's grandfather later purchased that property as well as the house at 110 High Street, (now 320 Market Street). That was Edward's boyhood home with his father and stepmother, Hannah, when he was not at Benfield. The house still stands as the Franklin Court Printing Office run by Independence National Historical Park, although Franklin's actual print shop was located nearby. Figure 1.1 in Chapter One shows these houses in 1723.

Edward's grandfather Benjamin owned far more real estate. As noted previously on page 10, in 1741, his heirs received 1,183 acres of land, plus eleven additional properties. The estate inventory of personal goods was valued above £4,400; more than £4,000 in bonds, bills and mortgages, almost £400 in cash and wearing apparel, and £65 in silver plate.

Benjamin left Edward a fifth of his silver, among which may have been a silver porringer now in the collection of the Metropolitan Museum of Art.[314] It was donated by Judge A.T. Clearwater who claimed that it was obtained from one of Edward's lineal descendants. The Judge went on to assert that Edward used this vessel to serve a terrapin stew flavored with Madeira from his famous cellar.[315]

Edward also inherited a third of the estate residue not itemized, as well as extensive property holdings including 740 acres in the Manor that were willed directly to him and contained his Benfield residence and farm.

In 1753, with Edward established nearby and his parents deceased, the Duffield family sold 320 Market Street to wigmaker George Cunningham, who as already noted was Franklin's barber. Cunningham paid £350 to Joseph's heirs with Edward taking a £100 mortgage for his share. A July 7, 1759, fire insurance survey was made for George's widow (Figure 8.1). The document concisely described Edward's boyhood home.

A general land tax passed in 1755 to fund a colonial militia. The tax was enacted directly after the Paxton Boys crisis. Handwritten tax records offer key information gathered by assessors, who noted current owners and prior landowners still receiving ground rent. If the resident was not the owner, assessors recorded occupation, race if not white, building value, and number of servants, enslaved, horses, and cows. Hannah Benner Roach compiled values of the city's taxable property from this list arranged alphabetically within each ward.[316] Edward's city property was assessed at £50, not reflecting his county holdings.

Horologists with similar property values included Henry Flower (£40), John Wood, Sr. (£30), Joseph Wills (£40), and Thomas Stretch (£60). Edward's fellow gentlemen taxpayers showed substantially higher valuations: Jacob Duché (£100), Dr. John Kearsley (£140), and Thomas Say (£130). Merchants included John Wister (£180), Samuel Mifflin (£240), John Mifflin (£150), Joseph Richardson (£120), and James Logan (£120). Carpenter-architect Samuel Rhoads was prosperous (£100) as was Joseph Fox (£220). At the tip of the economic pyramid were those identified as "Esq.," including William Plumstead (£250) and William Allen, perhaps the city's richest man (£600). Edward associated with all these men.

A 1756 war tax was collected for the next twenty years, taking one-and-a-half shillings per pound of assessed property values. This more than tripled tax burdens on householders paying provincial, county and city excises. Tax records provide details about property holders, including Edward (incorrectly called a brick-maker) who was assessed at £50 in the Mulberry Ward and £45 in the North Ward.

Edward's own home and adjoining building were evaluated by a September 4, 1758, fire insurance survey. Both structures are long gone and modern buildings occupy the corner. Joseph Fox did the appraisal but did not mention a shop, work space, or exterior clock:

Surveyed 4 Sep. 1758 Edward Duffield house situate at North Side of Mulberry Street at the corner of Second & Mulberry Streets, find 25 Foot on Second & 34 foot on Mulberry Street, 2 Stories, a [...] Wall on the South, d.o. on the North to the Garrett floor, then four to the top, Board Newell Stairs, plaistered Partitions 2 Stories painted. Shingling about 5 yrs. Old. Also, his house adjoining on the West [...] 18 foot 3 Inches on Mulberry including a three foot alley and 13 foot 3 Inches back, 2 Stories high. 9 Inch Wall, plaistered Partitions, Board Newell Stairs, Shingling about 5 yrs. Jos. Fox £350 @ 25/Ct. Back 28 £150 @25/Ct.[317]

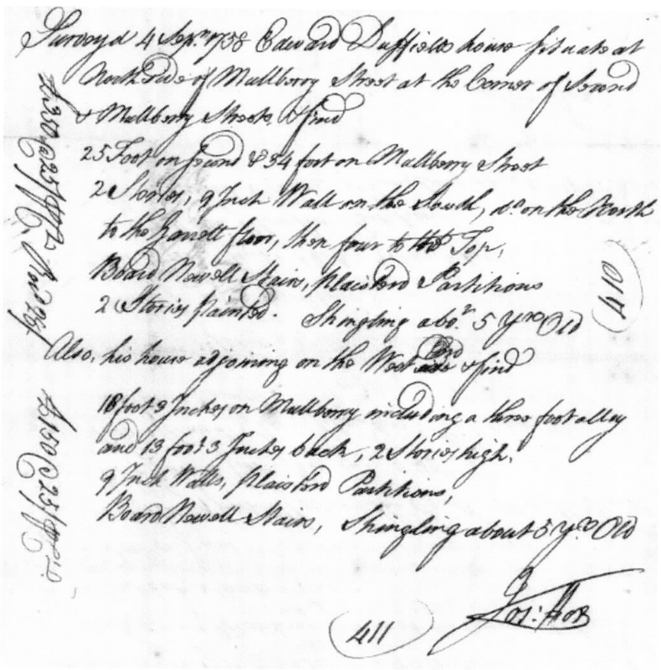
Surveyd 4 Sepr 1758 Edward Duffields house situate at
North Side of Mulberry Street at the Corner of Second
& Mulberry Streets & find
25 Foot on Second & 34 foot on Mulberry Street
2 Stories, 9 Inch Wall on the South, do on the North
to the Garrett floor, then four to the Top,
Board Newell Stairs, plaistered Partitions
2 Stories painted. Shingling abot 5 yrs Old
Also, his house adjoining on the West side & find
18 foot 3 Inches on Mulberry including a three foot alley
and 13 foot 3 Inches back, 2 Stories high.
9 Inch Walls, plaistered Partitions,
Board Newell Stairs, Shingling about 5 yrs Old
(410)
(411) Jos: Fox

Figure 8.2 1758 survey of Edward Duffield's home, Insurance Survey S00410, courtesy of the Philadelphia Contributionship for the Insurance of Houses from Loss by Fire.

In October 1764, Edward was selling property: To be sold by publick Vendue, at the London Coffee-House.... a Lot of Ground, with two Wooden Tenements thereon, that Lett for Fourteen Pounds a Year, situate on the North Side of Keys's Alley... subject to a proportionable Part of a Ground-rent of Four Pounds a Year....[318]

Nova Scotia Governor-in-Chief Montague Wilmot offered 100,000 acres of new land grants in 1765.[319] Speculators included Edward, Benjamin Franklin, Joseph Richardson, and twenty others. Franklin later wrote that he envisioned this land benefitting his grandchildren, but those Canadian deeds were void and worthless after the Revolution.

In February 1767, a new poor law was enacted. Its recorded assessments again provide data on individual citizens. Property was taxed to fund an almshouse and workhouse for destitute residents, including discharged soldiers unable to find employment. Cabinetmaker William Savery, a likely maker of clock cases, was an overseer of the poor. Edward was assessed for his house, horse, and two enslaved workers (Figure 8.3). This is the first record of Edward owning "Negros" although two were bequeathed to others in his father's will and may have been in his boyhood home.

In July of 1767 Edward added to his property holdings, purchasing 849 acres in three tracts of land in Cumberland County, appealingly named Walnut Hill, Vinyard, and The Rich Bottom.[320]

North Ward tax records in 1769 again show Edward with a horse and two enslaved Blacks. He paid a substantial tax of £156.19.4 on his properties. As a contrast, William Jones in Southwark was a tenant who paid £7 rent, while Clockmaker Jacob Godshalk (Gotshalk) had a single horse and servant on which he paid a total of £20.1.4 for the so-called Proprietary Tax. Five years later, in 1774, he paid a Provincial Tax of £12.9.0 with no land, animals or servants noted. He, but not Edward, was identified as a clockmaker perhaps indicating that Edward was better known as a landowner in this period. Edward was among the richest ten percent who possessed nearly seventy percent of the property and received almost ninety percent of rental income in a city of concentrated wealth.

**Figure 8.3** Detail of 1767 property tax assessment for Edward Duffield from assessor's handwritten page. Courtesy of Torben Jenk.

The 1772 Mulberry Ward tax list showed Edward paying a substantial £160.7.4 on his dwelling, enslaved worker, horse, and rents. Ground rents came from ten named individuals, from unnamed free Blacks, and for twenty acres of land. In February of the next year, Edward purchased two acres of land in Frankford from Rudolph Ness and his wife.

Mulberry Ward tax records in 1774 show Edward owning 150 acres, far more than nearly any other taxpayer in this densely populated neighborhood. He paid £192.18.7 on his property that included a horse and servant, and on rental properties occupied by several named tenants. He also purchased 300 acres in Lancaster County.

David Rittenhouse, who moved into town from his rural family home, was listed as an instrument maker (Edward was never listed as such) in the North Ward. He was taxed as a renter with no horse, cattle, or servants. In comparison, Edward paid taxes on rented dwellings in the Mulberry, South, Middle, and Lower Delaware wards. Just nineteen percent of the city's families owned their own houses, the remainder rented from landlords such as Edward. Wealth remained concentrated; the top ten percent of property owners accounted for eighty-nine percent of assessed values.

Edward owned fifty undeveloped acres in Bucks County Bensalem Township and in 1778 paid a tax of £1.10 on them.

Listed in 1779 as a farmer (not clockmaker) four years after his move from the city, he paid £50 for the Five Shilling State Tax, and £10.15 was paid for the same tax by his tenant, Lewis Brahl. His taxes in Moreland, based on a £18,700 assessment, were £280.10 and were among the highest in the area.

He paid £334, far above the average, for the new Federal Effective Supply Tax enacted by the Continental Congress to pay and supply its beleaguered soldiers. "Federal" signified an agreement by the thirteen colonies, but each could decide how, and if, to collect the tax.

The next year, he paid even more, £477.10, on his Philadelphia County property that was assessed at £19,100, and he paid £12 on a £4,000 assessment in Northern Liberties. Some of his tenants and property purchasers were listed on the tax rolls: Jacob Cooper, printer John George, merchant Luke Keating, smith Lewis Brahl, and skinner Martin Worknot.

In the 1781–1783 period, Edward paid £21.1.4 in Effective Supply Tax on a valuation of £3,371. He paid another £5.19.6 to the city for a £400 property valuation, and another £3.2.3 on £485 in value. The rolls showed tax payments relating to his tenants Martin Worknot, Hannah Wharton, Mrs. Bogarth, and Widow Martha Levi. Another taxpaying tenant, James Withy, was on Chestnut Street and advertised "fashionable Beaver and Castor HATS of the best quality…"[321]

His federal tax in 1782 was £3.2.3 on a £485 valuation. Edward's city property was worth £400, taxing him £2.1.8. His tenant printer, John George, paid £1.2.11 on a £220 valuation, and other tenants in the South and Mulberry Wards included Christina Bolitha, merchant Luke Keating & Comp'y, Martha Levi, and Mary Wharton.

The 1782 assessment of Edward's Moreland Township farm totaled £3371. There were 500 acres at £2750, two horses £36, six cows £25, plates (silver) £30.16, three "Negroes" £140, eleven sheep £3.6, one coach £60, one chair (rolling chaise) £16, and "Occupation" £300. Edward owned both a valuable coach and a chaise that he used for local travel and drives into the city. Two decades later, an "Old chair 2 wheels" was listed in his estate inventory.

The next year, he again was taxed in Moreland for his coach, as well as for properties in Northern Liberties, Mulberry Ward, and Bucks County. Federal tax was paid for ground rent of properties still occupied by Widow Levi and hatter James Withy. A Philadelphia County assessment listed his 500 acres, two horses, six cattle, sixteen sheep, and three enslaved Blacks.

Edward's rental properties at this time included a typical small house at the corner of Bread Street and Quary Street, or Moravian Alley, that was surveyed in 1784 by the Contributionship. It was two stories high, had twenty-four feet of frontage, and a depth of nine feet, eight inches.[322]

That year In Moreland, Edward again was taxed £7.10 for his coach and another £1.10 for his horse-drawn chaise. More land was purchased; the State Land Office issued Edward a warrant for a 400-acre tract in Pennsylvania's Luzerne County.

Intriguingly, Edward paid taxes on the former home of the late city leader James Logan. Tax records in 1785 for Second Street show an assessed value of £100 and tax of £8.8 for the Logan Estate. An £80 valuation appears the next year, referencing the west side of Second Street going south between Chestnut and Walnut streets. We do not know if Edward rented or at some time resided at this historic property.

The 1786 tax records for the Middle and Walnut wards showed Edward's properties on Third Street, Green Street, and for Allen's and Widow Levy's ground rents, as well as for James Logan's former dwelling. He paid taxes on land in Byberry, Moreland, Northern Liberties, and Bucks County.

Unlike Edward's prior large tax assessment for his unspecified "occupation," in 1787, the amount was just £19. Perhaps he had more officially retired by then. For his Benfield estate, he was taxed on 500 acres, his dwelling and tanning yard, five horses, six cows, one coach, one chair (chaise), four enslaved Blacks, seventy-seven ounces of silver plate, and one servant, totaling £2,915. Other taxed properties were in the North, South, and Mulberry Wards, and in Byberry, Moyamensing, and Bensalem Townships.

About that tanning yard and according to Martindale, Edward "…had taken a load of bark to Frankford. The tanner's prices were not satisfactory. Duffield therefore complained. 'If you do not like my prices,' said the tanner, 'you can go into the business yourself.' 'I will,' said Duffield, and going home, he established a tan yard."[323]

Tax records for 1788 showed shopkeeper James Paxton in an Edward-owned dwelling valued at £300, and Edward again was taxed on properties in the North and Mulberry Wards, Bensalem Township, ninety acres in the County of Huntingdon, and for his coach and chaise.

Martindale noted Edward's innovative stone barn: "Edward Duffield built a barn in Moreland, near the village of Somerton."[324] Along with a barn built by John P. Townsend in Mechanicsville,

"These were the first barns with stabling underneath and a bridge to get to the floor and mows. They are generally known as cellar barns and have been the fashion ever since." A detailed tax assessment from 1797 recorded the barn's dimensions and valued it at more than twice Edward's stone house.

Edward's barn disappeared long ago, but similar stone examples remain. One built in 1775 by John Bartram, Jr., still stands at Philadelphia's Bartram Gardens. Another from 1787 is at James Logan's Stenton estate.

When in 1803, Dr. James Mease published his five-volume American edition of Dr. A.F.M. Willich's *Domestic Encyclopedia…*, he described Edward's barn in great detail and included an illustration of Edward's improved barn-door hinge:

> The first barn built in the northern townships of Philadelphia county, upon the very excellent German plan, was by Edward Duffield, Esq. in the year 1789, and since that period his model has been copied by many in the neighborhood.
>
> The following particulars respecting it, deserve notice from all economical farmers.
>
> There are no door frames, except to the main door opening on the threshing floor. They are hung upon hooks and eyes, but in a different manner form the common mode, the hooks being welded to the hinge, and inserted into the eye, which is let in between two stones in the jamb of the door-way, Thus (illustration).
>
> The advantage of this plan is, that if the hook breaks, it is much more easily supplied by a new one when welded to the hinge, than when the hook is in the wall and the eye in the hinge.
>
> The threshing floor is let into a rabbet, in the cross girder. The rabbet is about five inches deep. The partition between the mows are scribed down to the floor, and nailed to the rabbet, and thus the leaking of the grain, which is so seriously complained of by most farmers, is prevented.
>
> The floor is of white oak plan, well seasoned and the boards broad; both edges are grooved and put together with tongues. The sill of the threshing floor is a long stone, and the upright post of the frame rests upon two stones which join the one forming the sill. By this arrangement there is little inequality in the rise from the bank, and the wheels of the cart or wagon go over easily.
>
> The joists are of sapling hickory trees cut and barked in the spring. Mr. D. says, such joists are more durable and strong than any other which can be put in a barn; and that trees cut and barked in the spring escape attacks from the wood worm, whereas if the bark be permitted to remain on a felled tree, it will inevitably be attacked in the course of the season. Mr. D's barn is 61 by 37 feet; but he observed that the nearer a barn approached to a square, the more economically it might be built.[325]

**Figure 8.4** Edward Duffield newspaper advertisement selling a plantation, *Aurora Advertiser*, December 20, 1791, p.1. Courtesy of Newspapers.com.

The first Constitution-mandated national census, gathered in 1790, recorded Edward at the Manor of Moreland, and in 1791 Edward placed a front-page ad in December indicating that he was selling some property adjacent to his home (Figure 8.4).

The property described in this chapter's heading sold at auction for £3,500 in April 1796 to Captain Stephen Decatur, now Edward's neighbor, who was a naval officer during the American Revolution and famously commanded the American privateers *Fair American* and *Royal Louis* that captured many British prizes at sea.[326] The bill of sale was signed by both Edward and his son Edward, Jr.[327] Decatur's son, Stephen, Jr., (1779–1820) was an even more famous naval officer and commodore.

DUFFIELD HOMESTEAD, BENFIELD.

**Figure 8.5** Photograph of Duffield Homestead, Benfield, *A History of the Townships of Byberry and Moreland, in Philadelphia, Pa,* by Joseph C. Martindale and Albert W. Dudley, 1901, p.303. Courtesy of Internet Archive.

On October 1, 1798, an assessor in Edward's area completed an official Federal form with the extremely lengthy title:

A List or Description of all Lands, Lots, Buildings, and Wharves, owned, possessed, or occupied, on the 1st Day of October, 1798, in Moreland Township, being within the 5th Assessment District & 1st Division in the State of Pennsylvania excepting only such Dwelling-houses, as with the Out-houses appurtenant thereto, and the Lots on which they are erected, not exceeding two Acres in any Case, are above the Value of One Hundred Dollars.[328]

Edward was number thirteen on the list, named as Occupant and Owner, with Silas Walton shown in the "Situation and adjoining Proprietors" column. The assessor recorded the following buildings and their dimensions: "1 Grand & Car.g house stone 30.20; 1 Corn House frame 20.21; 1 Cyder house frame 21.30; 1 Barn Stone 61.37; 1 Do Sawed Logs 38.18; 1 Tan House Stone 33.21; 1 Brew house frame 15.15." The total valuation was $6,318 with the stone barn still the most valuable at $750, followed by the house at $350.

An accompanying form added more specifics. It had a similarly expansive title: Particular List of Description of each Dwelling-house, which, with the Out-houses appurtenant thereto, and the Lot on which the same are erected, not exceeding two Acres in any case, were owned, possessed, or occupied on the 1st Day of October, 1798, in Moreland Township being within the 5th Assessment District and 1st Division in the State of Pennsylvania and exceeding in Value the Sum of One Hundred Dollars.[329]

The second form documented a two-story stone house thirty-six feet by twenty feet, a two-story stone kitchen thirty-two feet by twenty feet, a stone wash house eleven-and-a-half feet by twenty-feet, a twelve-foot by twenty-foot wood-frame brew house, a stone smoke house ten feet square, and a stone spring house twelve feet square. Assessed total was $2,000. A shop or building related to clockmaking was not mentioned.

Next year's form was signed by Edward's son at the bottom, with no apparent conflict of interest: "I the Subscriber Assistant Appraiser have made the foregoing valuation as Witness my hand this—day of—1799. Edward Duffield Junr."[330] It described the same buildings but added counts of windows and their many old-style small glass panes. Edward's two-story house had nine windows with 135 panes. The two-story kitchen featured thirteen windows with 138 panes. The single-story wash house had just two windows with ten panes. Other outbuildings had no windows.

Just one image of Benfield is known (Figure 8.5). The old photograph appeared in Martindale's book and showed the stone house and kitchen built against it. Christ the King Catholic Church, built in 1963, now occupies the site at 3252 Chesterfield Road in the Torresdale area of Philadelphia.

Edward's longtime Comly neighbors were listed above him on the form. Sarah Hepburn, Edward's remaining widowed daughter, was below, followed by John Foster and Jacob Overduff, all with Edward as the property owner. Sarah had a two-story stone dwelling forty feet by twenty feet on a half-acre, with nine windows of 123 panes, and a windowless smoke house ten-feet square. Mister Foster occupied a one-story log dwelling twenty-five feet by eighteen feet on an acre of land, with four windows of twenty-seven panes. Mister Overduff's stone-and-frame dwelling was fifty-four feet by twenty feet, with eleven windows of ninety panes. It sat on a half-acre with an additional one-story building, eighteen-feet square with three windows of fifty-four panes, and another ten-foot-square windowless stone smoke house. The total valuation of Edward's holdings at this location was $3,540. Other relations, John Swift and Jacob Duffield, owned properties close by.

Edward and his household again were included in the 1800 Federal census. William Lardner, assistant to the district's marshal, reported eleven unnamed residents, including a free white male aged ten to sixteen, two aged sixteen to twenty-six, one aged twenty-six to forty-five, and one, no doubt Edward, older than forty-five. There was one free white female between ten and sixteen and another between twenty-six and forty-five. There were two others, probably servants, and two enslaved Blacks. Except for Edward and Edward, Jr., we have no idea who the others might have been. Edward's widowed daughter, Sarah Hepburn, lived nearby in a smaller five-person household.

A small booklet clearly belonging to Edward is in the collection of the American Antiquarian Society in Worcester, Massachusetts. It is a *McCulloch's Pocket Almanac for the Year 1801* printed in Philadelphia.[331] Edward signed his name on the front cover and wrote on inside pages, mostly about making barrel staves, which farmers and merchants needed to manufacture barrels used to ship goods by wagons and boats (Figures 8.6, 8.7). On the almanac's pages, Edward also made notes about his tenant and employee, Martin Somer: "*May 4, 5, 6 1/4 till 1/2 past 8 rain, 7 (?) rain, 8, 11,12 5 3/4 days a 5 p day See Book.*" (This page had a large X over the writing). Edward signed his name twice on other pages and wrote:

> Feby 21 1801 David Snyder engaged to take all the staves I am now getting made at six pounds a thousand in the woods where they are made & where they are to be culled. But he will be content with eight or ten thousand if any other person will take the remainder on the same terms. E Duffield ….. March 9 1801 David Snyder engaged to take my Indian Corn 500 Bushl more or less at 6/9 certain p bushl he is to have 60 days But is to pay the

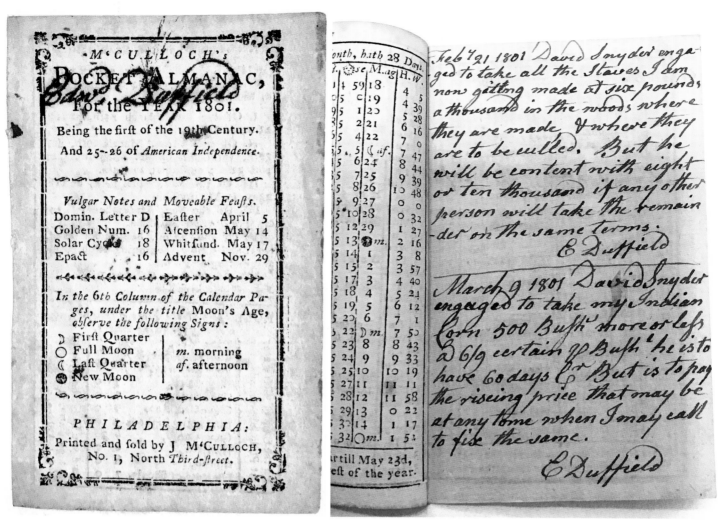

**Figures 8.6 & 8.7**  McCulloch's Pocket Almanac for the Year 1801. Courtesy of the American Antiquarian Society.

riseing price that may be at any time when I may call to rise the same. E Duffield…. Decr 2, 1801 Wm. Walton Jnr & his Brother began to make staves on the following terms rising If I sell then at £5 or upwards he is to have 25/arm for making but if I sell then under £5 p thousand he is to have only 3 Doll.s for making. The above prices is to be for them in the woods where made after culling…. Decr 4, 1801 Martin Somer agreed to make Staves on the same terms as above.

Edward was still buying land just a few months before he died in July 1803. He added 400 acres from Josiah Lewis in Northumberland County to his possessions. Meanwhile his son in April was selling an eighteen-acre lot in Moreland Township, with additional acreage available. Edward Jr. advertised a "handsome stone dwelling, with barn and stables…. about 18 miles from the city, on the Southampton road… an orchard, meadow, and wood land."[332] Edward's lengthy will, transcribed in Appendix 1, detailed lands and properties bequeathed to his heirs. From his extensive Benfield estate and its adjoining farms and buildings, to lands he owned in New Jersey, and to the substantial amounts of cash, silver, textiles, furnishings, and tools enumerated in his estate inventory, it is clear that Edward died a wealthy man, leaving his surviving family members comfortably situated.

CHAPTER NINE

# Edward Duffield and Benjamin Franklin

*Hide not your talents, they for use were made, What's a sundial in the shade?*
*You will find the key to success under the alarm clock.*
*A Brother may not be a Friend, but a Friend will always be a Brother.*
— Benjamin Franklin

Nearly every mention of Edward Duffield attaches great importance to his friendship with Benjamin Franklin (Figure 9.1), twenty-four years his senior, and his serving as an executor of Franklin's will. This chapter explores the Franklin-Duffield connections in greater detail than ever before. Certainly, the two men and their families were close for decades and often spent days and weeks in each other's company. Many letters by Franklin and his family mention Edward and his family and send them regards. But there are no known letters between the two men. More than two centuries of scholarly research into Franklin's life, and dozens of published volumes of Franklin correspondence, should have uncovered at least one if written.

One possible explanation is that William Temple Franklin (1760–1823) had charge of thousands of Franklin papers after his grandfather's death, and his carelessness with them led to troves being destroyed and given away. Franklin's will directed Temple to collect, edit, and publish the papers, but the grandson dallied for decades. Some papers were discovered by historian Jared Sparks in 1831 in London, where a tailor was using them for sleeve patterns. Sparks found more in a Philadelphia stable, where visitors were given handfuls as souvenirs. So Franklin's letters to Edward may simply have been lost. But at the same time, it is unlikely that Edward's direct descendants would have discarded or misplaced letters by the great man. Edward, Jr., lived with his father and remained a sophisticated gentleman at Benfield until his own death. Further descendants were equally erudite. If Franklin letters existed and were cherished by later generations, nobody yet has brought them to light as might be expected. So perhaps the two men never corresponded despite the close friendship between them.

There was no need for Edward and Franklin to correspond when living nearby and conversing in person. But during the decades when Franklin resided in England and France—he spent only three years on American soil between 1757 and 1785—we might expect there

**Figure 9.1** 1767 Portrait of Benjamin Franklin by David Martin. Courtesy of the Pennsylvania Academy of Fine Arts, 1943.16.1

to have been an exchange of letters between the two men as Edward and Franklin shared a common interest in horology and Franklin might well have had information he wished to share directly with Edward. The years 1757 to 1775 were the primary period of Edward's horological career, when Franklin was purchasing clocks and watches in England and engaging with prominent English horologists.

Eminent English clockmaker John Whitehurst crafted clocks for Franklin and hosted the American at his Derby home. A Whitehurst-signed example of a Franklin three-wheel clock has been examined by the author, and another hangs at the Derby Museum. Franklin's purchases and letters from Whitehurst are discussed and transcribed below. Franklin also employed the renowned English horologist, John Ellicott, and was acquainted with James Ferguson, the eminent Scottish astronomer, horologist, scientist, and author, who in 1758 adapted Franklin's three-wheel clock design. Ferguson's improved version ran a full week, rather than one day, between windings and better indicated the time, which was confusing on Franklin's original dial.[333]

One could imagine Franklin and Edward directly exchanging letters about Whitehurst, Ellicott, Ferguson, and Harrison, but no such letters exist. In 1767, Franklin did ship to Edward a published treatise by horologist John Harrison, of longitude fame, whose London workshop he visited. The book came to Edward indirectly, however, via Deborah Franklin. She also later relayed Edward's thanks back to Benjamin.

Benjamin Franklin was born in Boston, and in October 1723, at age seventeen, he arrived in Philadelphia as an apprentice running away from his older brother's printing business. On his first day in the city, he spotted Deborah Read, his future wife, as he walked past her home at 108 Market Street. He returned later in the day and was accepted as a boarder in a room above the family's dry goods shop. The Duffield home was next door at 110. The families became friendly well before Edward's birth seven years later.

On October 11, 1726, Franklin returned to Philadelphia from his first stay in London and made Deborah his common-law wife. Her first husband, a bigamist named John Rogers, had fled the city but had not been declared legally dead.

The next year Franklin formed his Junto whose members, all artisans, craftsmen, and petty merchants, wore durable leather aprons as they labored. The original twelve members discussed questions posed to them in advance of their evening meetings, and they met monthly on Sundays for calisthenics. Many remained Franklin's friends and later were Edward's.

In 1728, Franklin began publishing the *Pennsylvania Gazette*, buying it from Samuel Keimer. It became the most popular newspaper in the American colonies partly due to Franklin's management and control of the postal system. When in 1748 he retired in his forties from the printing trade, he left management of the profitable business to his English immigrant partner David Hall. Perhaps his early retirement spurred Edward to do the same at age forty-five in 1775.

Franklin's illegitimate son, William, was born February 22, 1730, shortly before Edward in April. His birth mother was a secret that Franklin never revealed. The lad was raised by Deborah, whom he always called Mother. The two boys must have been acquainted from early in life. William was tutored by Professor Theophilus Grew on Arch Street and attended an academy run by Alexander Annand.

Franklin in 1732 began publishing *Poor Richard's Almanac*, which included equation-of-time tables for regulating mechanical timepieces with sundials. Other timekeeping information appeared as well. Under his pseudonym Anthony Afterwit, Franklin wrote humorously about clock ownership:

> It happened frequently, that when I came home at One, the Dinner was but just put in the Pot; for, My Dear thought really it had been but Eleven: At other Times when I came at the same Hour, She wondered I would stay so long, for Dinner was ready and had waited

for me these two Hours. These Irregularities, occasioned by mistaking the Time, convinced me, that it was absolutely necessary to buy a Clock; which my Spouse observ'd, was a great Ornament to the Room![334]

During the same year, 1736, when Edward lost his mother and four siblings, Franklin's beloved four-year-old son, Francis, or Frankie as he was called, was one of the city's 158 smallpox casualties. After his death, Franklin rebutted false claims that the boy died from being inoculated. Franklin revealed that he had postponed the controversial procedure until his son recovered sufficient strength from a long bout of the flux. The Franklins had one other child together, their daughter Sarah, or Sally, as she was known.

In 1741 when Edward was age eleven, Franklin printed a fifty-six-page catalogue of 375 books owned by the Library Company of Philadelphia. Franklin founded the membership library, still thriving today, ten years previously. The titles represented practical history, literature, and science, unlike most libraries of the time which focused upon theology and classics. Young Edward may have used this educational resource.

In 1743 Franklin collaborated with botanist John Bartram to issue a lengthy proposal leading to the birth of the American Philosophical Society. The two men stated: "The first drudgery of settling new colonies, which confines the attention of people to mere necessaries is now pretty well over; and there are many in every province in circumstances, that set them at ease, and afford leisure to cultivate the finer arts, and improve the common stock of knowledge."[335] Despite Franklin's preliminary exertions, the APS was dormant for the next two decades until reactivated in the 1760s. Edward was elected a member in 1768.

During a family visit to Boston in 1743, Franklin met Dr. Archibald Spencer and witnessed some exciting electricity demonstrations. These sparked Franklin's famous experiments conducted in close collaboration with Kinnersley, who later traveled throughout Colonial America demonstrating and lecturing on their discoveries. Teenage Edward may well have observed and participated in the literally hair-raising experiments. His great-grandson Edward Duffield Neill claimed that Edward "was constantly devising philosophical apparatus for his brother-in-law, Prof. Kennersley [sic], and for Dr. Franklin."[336]

Aged forty-two in 1748, Franklin sold his printing business, hung up his leather apron, and devoted himself to more gentlemanly pursuits including his electricity experiments and civic activism. Despite his persistent advocacy of diligent labor, he abandoned his trade at a relatively young age, when he could afford to stop dirtying his hands. Other craftsmen later did the same, including Edward and cabinetmaker Jonathan Gostelowe, mentioned earlier, who married into the Duffield extended family.

From 1748 to 1753, Franklin paid the annual ground rent of £3.1.4 for property at today's 131 Market Street, where he and his family lived from 1738 to 1748. The money was paid to the estate of Benjamin Duffield, Edward's grandfather. Thomas Whitton, one of the executors, signed the receipts. A 1753 payment for this amount is recorded in Franklin's ledger as paid to Jacob Duffield, Edward's cousin, a son of his father's brother Thomas. This further illustrates the ongoing connections between Franklin and the Duffield family.

"The Franklin Ledgers Database,"[337] the result of a recent multi-year digitizing and transcribing project at the APS, now offers online access to the voluminous account books maintained by Benjamin and Deborah in their shop. Within this wealth of information are records relating to Franklin's position as Postmaster in Philadelphia. A search for Edward's name revealed scans and detailed summaries of nine entries, all for incoming postal charges. At this time, recipients, not senders, of mail paid the postage depending on distance the mail traveled. The first five lines simply listed Edward's payments of £0.1.7 or £0.3.3 in 1752 and 1753. The remaining four lines, recorded in February, April, May, and July of 1752, were for the same apparently standard charges but added that two originated in Boston and two were from New York. The Boston packages traveled farther and perhaps were heavier, at "7.16," compared to those from New York at "3.16." We can imagine young Edward entering the Franklin shop to collect his mail, but speculating is impossible on who sent it and what it contained. We know of no personal or commercial connections Edward had to those two other colonial cities.

Franklin published at least two accounts in the 1750s that demonstrate his interest in clocks and timekeeping devices, an interest obviously shared with Edward, and it is not hard to imagine the two men discussing the events described by Franklin in these accounts. In 1752, Franklin published an account of lightning striking two houses on Society Hill, writing that the bolt "… seem'd to go considerably out of a direct Course, for the sake of passing thro' Metal; such as Hinges, Sash Weights, Iron Rods, the Pendulum of a Clock, &c…."[338] In Franklin's 1757 almanac he humorously proposed an oversize sundial "by which not only a Man's own Family, but all his Neighbours for ten Miles round, may know what o'Clock it is, when the Sun shines, without seeing the Dial."[339] He suggested an array of twelve large magnifying glasses focused onto a semicircle of seventy-eight cannons—massive thirty-two pounders recommended—that would sequentially fire those guns every hour as the sun moved across the sky. We can easily imagine that Edward was amused.

Franklin wrote his first will dated April 28, 1757, prior to his next departure for England. He went as emissary for the Pennsylvania Assembly in its dispute with the colony's proprietors. His son, William, and two slaves, Peter and King, accompanied him. Edward was named an executor of that initial will as well as of the final version in 1790.

When in London, Franklin visited John Harrison and paid to view a longitude clock.[340] He would have examined a large, complicated timepiece that proved sufficiently accurate to determine longitude at sea. Harrison had not yet completed his "H4" that belatedly earned him a very large monetary award from the British government under the terms of the 1714 Longitude Act. This visit was something that would assuredly have interested Edward, and was one of the moments when we might have hoped for a letter from Franklin to him.

John Waring wrote to Franklin in 1757 to introduce the Bray Associates' mission to educate Black children and instill their Christian faith. The Bray School was another way that Franklin and Edward shared common interests as both men were involved in its long history. Franklin would have known of Thomas Bray, D.D. (1656/8–1730) from the zealous Englishman's previous donation of many religious books to Christ Church, as part of Bray's philanthropy to create parish libraries in Colonial America. A Philadelphia school was opened by the Bray Associates under the direction of the Reverend William Sturgeon in November 1758.

Franklin was elected a member, and then chairman, of the Bray Associates in 1760. At his first meeting on January 17, he recommended that three additional American schools be established in Newport, New York, and Williamsburg. Franklin and British lexicographer Samuel Johnson attended a May 1 meeting (their only known encounter), at which time a letter from Sturgeon was read. He reported that his school was proving successful with eleven boys and twenty-four girls enrolled.

The school continued to thrive, and in 1766, Edward and Francis Hopkinson took over the supervision of the school from Sturgeon, a role they continued until the fall of 1774 when the job was handed over to the Reverend Thomas Coombe, an assistant minister at Edward's church. The school continued its good works until March, 1845, when it was supplanted by the city's public education system. A complete history, "The Philadelphia Bray Schools: A Story of Black Education in Early America, 1758–1845," appeared in the October, 2023, issue of *The Pennsylvania Magazine of History and Biography*, researched and written by Grant Stanton and John C. Van Horne.

Claims that Edward was Franklin's clockmaker are overstated, and available evidence does not support this particular connection between the two men. We know that while in England, Franklin sent a letter to his wife mentioning an unnamed clock in their home: "In my Room, on the Folio Shelf, Between the Clock and our bed Chamber, and not far from the Clock, stands a Folio call'd the Gardener's Dictionary, by P. Miller."[341] But there is no evidence that this clock was made by Edward. Another letter noted that Franklin was shipping to her for Mr. Schlatter a box with a watch,[342] which Franklin had paid a London watchmaker to repair. Michael Schlatter, a Philadelphia German-American cleric and educator, obviously chose not to have his watch repaired by Edward or another of the city's watchmakers, something that Franklin may otherwise have recommended.

Franklin's financial records also do not reveal any purchases from Edward for clocks or repairs. However, a Duffield clock (Catalogue No.4) was passed down through the family of Franklin's daughter. Other services and timekeepers by Edward to Franklin may have been freely given but there is no record of them. Thus, it seems more accurate to say that while the two men and their families were closely acquainted for nearly all the years that their lives intersected, Edward was not primarily responsible for the Franklin family's timekeepers. Instead, Edward was Franklin's friend, a basic fact that

Figure 9.2   Dial and hand of three-wheel clock designed by Franklin and fabricated by John Whitehurst. Private collection. Author photo.

**Figure 9.3**  Mrs. Richard Bache (Sarah Franklin), 1793 portrait by John Hoppner. Courtesy of Metropolitan Museum of Art, 01.20.

**Figure 9.4**  Richard Bache, 1792–1793 by John Hoppner, National Portrait Gallery, Smithsonian Institution; gift of Richard Bache Duane Family, NPG.2011.145.

underscores Edward's importance as far more than a colonial clockmaker. Their friendship is confirmed in a long letter from London to Deborah in 1758 concluding with love to various friends, including "Mr. and Mrs. Duffield."[343] This is the first of many instances of Franklin sending from overseas his warm salutations to them, until on November 1, 1762, he again returned home to Philadelphia and reportedly entertained friends, almost certainly including Edward, morning to night.[344]

Franklin wrote from Philadelphia to London in 1763 to complain to the official clockmaker to King George III, John Ellicott: "I am sorry I cannot give you an agreeable Account of the Performance of the Watch. The new Spring unfortunately broke soon after I left England. Since my coming here, the old one is put in again [perhaps by Edward]; but I have not yet accurately adjusted the Watch so as to bring it to keep time as well as it us'd to do in London."[345]

While at home in Philadelphia, Franklin corresponded with other English horologists, and he may have shared those letters with Edward. John Whitehurst informed Franklin about Parliament's and the King's praises of John Harrison's achievements and sent "most affectionate respects" from his wife and Erasmus Darwin.[346] Whitehurst was a member of the Lunar Society and a friend of its illustrious members including Darwin, Joseph Wright, Matthew Boulton, and Josiah Wedgwood. His home in Derby formerly had been the residence of John Flamsteed, the first Astronomer Royal. Franklin visited Whitehurst in Derby in 1759 when traveling to Scotland, and again in 1760, 1771, 1772, and 1774.

Franklin replied to Whitehurst: "I am glad to hear Mr. Harrison is like to obtain some handsome Encouragement. I have heard that Mr. Graham, (the famous Graham of your Trade) should say, Harrison deserv'd the Reward if it were only for his Improvements in Clockmaking: The Error of his Watch in the Voyage between Portsmouth and Portroyal in Jamaica, was it seems but 23 seconds of Time! A surprizing Exactness, if it holds."[347]

Whitehurst constructed prototypes of simplified three-wheel clocks that Franklin designed and hoped would enable colonial makers to save metal and fabrication time (Figure 9.2). James Ferguson also wrote about these. However, the unusual dial of Franklin's clock was difficult to decipher, and the invention did not become popular.

On November 7, 1764, Franklin departed for another long stay in England, escorted to his ship by a cavalcade of 300 of his friends, Edward almost certainly among them. In the same year, Kinnersley, now professor of English and Oratory at the College and Academy of Philadelphia, publicized his course on electrical experiments that he had developed with Franklin.

Franklin wrote to his wife the next year and promised to "look out for a Watch for Sally, as you desire, to bring with me (Figure 9.3). The Reason I did not think of it before, was your suffering her to wear yours, which you seldom use your self."[348] We do not know why Deborah asked him to purchase a watch in London when she could have had one from Edward. Perhaps like many colonials, she felt that a watch from England was more desirable, and perhaps was less costly without importation expenses, than one sold by the local watchmaker no matter how good a friend he was. In any case, it was not a question of money, as she had full charge of her domestic finances.

A letter from her contained details about their new house: "In the parlour… your timepiece stands in one corner, which is all wrong, I am told."[349] She provided no maker name or details, but she might have been referring to the non-striking Duffield-signed clock—Catalogue No.4—now at the APS after descending in the Bache family. Likewise, she did not mention who had told her it was all wrong but perhaps Edward advised her of a problem. If so, he should have been able to correct it.

In June 1767, Franklin wrote to Deborah about John Harrison's new treatise: "I send a Book on Mr. Harrison's Watch. Present it from me to our ingenious Friend Mr. Duffield, with my love to them and their Children."[350] Franklin used "ingenious" many times to describe men he admired. The book for Edward was most likely *The Principles of Mr. Harrison's Timekeeper with Plates of the Same*, issued that year by the Commissioners of the Longitude. Deborah replied with her typically unique spelling: "Mr. Duffell is very much plesed with his presente and beges you to excepte his beste thankes for it…"[351]

A February 1768 letter from Franklin to his wife[352] again sent his love to "Mr. and Mrs. Duffield," and another in December[353] asked that he be remembered respectfully to them.

Franklin wrote to his wife and to his son-in-law, Richard Bache (Figure 9.4), that he at last had sent Sally "a new Watch"[354] as he had promised three years previously. Certainly, one from Edward could have been provided more swiftly. A subsequent letter from Franklin to Deborah reacted to her request for a watch for herself:

> I shall, if I can afford it, send you another for your self: I say if I find I can afford it; for I understand the Balance of the Post Office Account which I must pay here, is greatly against me, owing to the large Sums you have received. I do not doubt your having applied them properly, and I only mention it, that if I do not send you a Watch, it will not be thro' Neglect or for want of Regard, but because I cannot spare the Cash…[355]

Again, there is no suggestion of purchasing a watch in Philadelphia, perhaps for similar financial reasons.

In 1771 Franklin in London heard from John Whitehurst about clocks fabricated for him:
> I sent the two Clocks on Tuesday last by Clark; to the Bell in Wood Street, Directed for you in Craven Street, the Strand. I hope they will please and Come Safe to hand. I have been So much out, I could not Engrave the Plates, and therefore Chose to Send them in the State they are In rather than keep them longer. Mr. Tompson Engraver in red lyon Street Clerken well, will do them in a day or two at any time, giving him proper Directions.

**Figure 9.5a/b**  Replica of "Fugio" dollar coin designed by Benjamin Franklin. Author collection and photos.

Please to give him a line when you Chuse he Should wait on you. I am Sir your Most Obedient Servant.[356]

A week later, Whitehurst wrote again with an invoice and mention of a tide-indicating clock designed by James Ferguson:
I received your favour, and have sent the Account of the Clocks underneath. The other Clock which you was so kind to order is in hand and will soon be compleated. I found it necessary to depart from Mr. Ferguson's Plans, for the sake of greater simplicity. The moons southing, and time of high water, ought to be as visible as the time pointed out by the Indexes. I believe you will agree with me in this alteration. I consulted Mr. Ferguson before I alterd his Designs.[357]

Ferguson devised clocks showing local tides, and two were made for Franklin that indicated Boston Harbor's rising and falling waters. No tide-indicating clocks made by Edward are known although they would have been marketable in Philadelphia since the Delaware River is tidal. There are, however, known Philadelphia-made tide-indicating clocks in a private collection, signed by Edward's contemporaries Joseph Wills (1700–1759) and William Stretch.

Detailed correspondence from Whitehurst about a tidal clock continued in 1772:
Perceiving you were not quite Sattisfied with the Account I gave you of the Clock I made for you, I have now made another with a round dial, The hour hand of which performs one revolution 24 hours. Concentric to that, is a hand to Shew the time of high and low water, the hours of flood and ebb. Under the foot of the hour 12, is wrote highwater, at 6 Low water, in Strong characters. I am in hopes Sir that the Simplicity of this method will please you. If you should approve this Clock, the Dial is circular and 13 Inches diameter. If the London Artist make a difficulty in Executing the case, I will send one along with the Clock. In the clock now made I've made a provisions for Seconds, therefore please to Say, whether you choose Seconds or not, as the whole is compleated, and going.[358]

Three months later, Whitehurst wrote again and indicated that he had built the wood case as well: I have this Day deliver'd your Clock to Mr. Clark, who Inn's at the Ball in Wood street and hope it will Arrive Safe and to your Sattisfaction. Please to unlock the door, and you will find the Screws, which fasten the Case in the Packing Case. The weights, Ball, Pulley &c. are Packed between bottom of the Clock and Packing cases. And please to draw the head of the Case off before you attempt to take the body out. I've made all the parts of the Case stronger than Common, for the better measurement of time—to prevent any Vibration by the action of the Pendulum.[359]

More than ninety tall-case clocks by Whitehurst are known, and at least three were sent to Franklin and perhaps were among those listed, but not named, in Franklin's estate.

On December 19, 1774, Deborah Franklin died in Philadelphia five days after a paralytic stroke. Although married to Benjamin for forty-four years, she had not seen her husband for a decade. He learned of her death while preparing finally to return home from London. She was buried in the cemetery of Christ Church near members of Edward's family. No doubt he and his family attended the funeral and burial of this lifelong family friend.

Before Franklin departed from England, he did a personal favor for Edward. In March of 1775, Franklin wrote to a friend in Edinburgh, the eminent Lord Kames, presenting Edward's son Benjamin, who then was studying medicine there.[360] This may relate to the abject letter later sent to Franklin in Paris when Benjamin was stuck in Bourdeaux, transcribed in Chapter One. Soon afterwards in May, Franklin returned to Philadelphia, having abandoned hopes for reconciliation with England. He brought with him a substantial amount of furniture and most likely the Whitehurst clocks.

In the politically tumultuous month of June 1776, Franklin stayed at Benfield with Edward while recovering from a gout attack. Franklin wrote to George Washington and to Dr. Benjamin Rush about his whereabouts, the "Manor of Moreland, at Mr. Duffield's."[361] Franklin was on the Committee of Five appointed by Congress to compose a declaration of independence. Their first meeting, with Edward as host and possible witness, was at Benfield, from where Jefferson returned to the city to continue work on the historic manifesto.

Perhaps with Edward's encouragement, Franklin used sundial imagery when designing the first Continental currency. The coin, and subsequent paper bills, included the words Fugio (Latin for "I flee") and "Mind Your Business" to remind his fellow citizens that they should stay busy and productive to support the economy and war effort. After the Revolution, the first copper coinage minted for the new nation, the so-called Fugio Cent, also incorporated Franklin's 1776 sundial design (Figures 9.5a, b).

It is certain that the Duffields often offered the hospitality of their Benfield Estate to Franklin's family and that this hospitality was returned. On August 17, 1776, when Franklin was back at his Philadelphia home, sixteen-year-old William Temple Franklin overnighted at Edward's and wrote to his grandfather: "It being rather late when I got to Mr. Duffields and the Road from there to Mr. Galloways being very bad; by the kind invitation of Mr. and Mrs. Duffield I staid that Night…. Let Mrs. Bache (Franklin's daughter) know that her Son William has been very well, except now and then the Musick in his Ear, but by Mrs, Duffields good nurseing is now much better."[362]

Young William apparently was under the Duffields' care while his parents dealt with the severe illness of his baby sister. She died on the very day that this letter was written.

After she fled her home in British-occupied Philadelphia, Sarah Franklin spent time with the Duffields in the fall of 1777 and wrote to her father about Edward's arranging for cloth that was in very short supply: "I can assure you, my dear papa, that industry in this house is by no means laid aside. Mr. Duffield has hired a weaver that lives in his farm to weave eighteen yards by making three of four shuttles for nothing and keeping it a secret from the country people, who will not suffer them to weave for those in town. I think myself lucky to have such a friend."[363]

An October letter from Isaac All to Franklin added: "Mrs. Beache [sic] was Just returned to Town from Mr. Duffelds where she had spent some time in the very warm weather and was busily employed in cutting out and making up shirts for the Soldiers of Gen. Washington's Army."[364]

A letter came from Franklin to Sarah: "Present my affectionate regards to all friends that enquire after me, especially Mr. Duffield and family...."[365] Sarah wrote to William Temple Franklin: "The Dufeild [sic] Family, and the Miss Cliftons I delivered your compliments too, and they desire to be remembered to you in a particular manner."[366] Sarah's husband Richard Bache wrote to Franklin, "Miss Sally Duffield is now with us, she sends Compliments to you & Temple."[367]

A 1779 letter that Franklin, now in France, received from his daughter, Sarah Bache, confirmed Edward's family's place in society, "Mr. Duffields Family desired when I wrote to remember them to you....The youngest daughter I have introduced this winter to the Assembly. She is like her mother. The Ambassador (French) told me he thought her a great acquisition to the Assembly."[368]

A July 1780 letter from Bache to Franklin reported that his wife "with her two youngest Children are in the Country, at our Friend Mr. Duffield's, returning a visit that the Miss Duffields made us in the Winter."[369] Her September letter elaborated on this:

> ... tis but a few days since I came from Mr. Duffields, where I have been with the two youngest Children all Summer, they are and have been perfectly well, tho it has been the most Sickly Summer ever known in town, & particularly fatal to young Children... the whole Family at Benfield were so kind and Affectionate, to me and the dear little Babies, that I shall ever feel the strongest attachment to them.[370]

Franklin's February 1781 letter to the son of a Paris neighbor, Louis Le Veillard, accompanied a package for the young man to carry to America. It contained seeds (not clocks or watches) for Edward which Franklin urged the Frenchman to keep dry.

In July Sarah wrote to her father: "The two youngest Children (Elizabeth and Louis) Mrs. Duffield took home with her last week, I heard yesterday they were well, we were a good deal loth to trouble the good Family with them again, but they insisted on it, and I am a great coward about keeping such young Children in this hot town during the warm weather."[371]

An ornate pocket watch in the collection of the Temple University Library was gifted to Franklin by English politician David Hartley (1732–1813) upon the signing of the 1783 Treaty of Paris.[372] Edward most likely saw this fine timepiece when Franklin finally returned home from London in 1785 and was greeted with great public acclaim. Early the next year on January 17, Franklin's birthday was celebrated at the Bunch of Grapes Tavern, where Edward likely was present for the festivities.

In 1786, Franklin's English clockmaker and Royal Society Fellow, John Whitehurst, became an overseas member of the APS per Franklin's recommendation. Whitehurst sent a letter of thanks:

> I take the earliest opportunity of returning my grateful acknowledgements for the great honour which the learned Philosophical Society of Philadelphia have been pleased to confer upon me, by chusing me as a member thereof. I shall at all time be extremely happy in contributing anything in my power in support of so laudable an Institution, but am afraid the best of my endeavors will avail little. However, as some testimony of my good intentions, I beg the favour of the Society to accept a copy of my Inquiry, Sent herwith, which I hope will soon be followed by a Copy of my Next work viz An attempt to obtain invariable Measure from the Mensuration of time &c. This work is now in the press and I hope will soon be forwarded...[373]

Reverend Manasseh Cutler, traveling from Hamilton, Massachusetts, recorded on several pages of his journal a visit to Franklin's home on Market Street in July 1787. Cutler was a distinguished scholar and botanist. Long diary extracts were reproduced in Jared Sparks' 1853 *The Life of*

*Benjamin Franklin* and presented "an interesting picture of Dr. Franklin's appearance and manner at this period of his life"[374] when Edward may often have been with the aged patriot:

> His voice was low, but his countenance open, frank, and pleasing… I was highly delighted with the extensive knowledge he appeared to have of every subject, the brightness of his memory, and clearness and vivacity of all his mental faculties, notwithstanding his age. His manners are perfectly easy, and every thing about him seems to diffuse an unrestrained freedom and happiness. He has an incessant vein of humor, accompanied with an uncommon vivacity, which seems as natural and involuntary as his breathing.

At that time, Franklin owned the largest number of books in private hands. Cutler penned a lengthy description of the home's library, where Edward no doubt sat many times, and its many curiosities:

> It is a very large chamber, and high-studded. The walls are covered with book-shelves, filled with books; besides there are four large alcoves, extending two thirds the length of the chamber, filled in the same manner… He showed us a glass machine for exhibiting the circulation of the blood in the arteries and veins of the human body… Another great curiosity was a rolling press, for taking the copies of letters or any other writing. A Sheet of paper is complete copied in less than two minutes; the copy as fair as the original, and without defacing it in the smallest degree. It is an invention of his own, extremely useful in many situations in life. He also showed us his long artificial arm and hand, for taking down and putting books up on high shelves, which are out of reach; and his great armed chair, with rockers, and a large fan placed over it, with which he fans himself, keeps off flies, etc. while he sits reading, with only a small motion of his foot…[375]

Perhaps Franklin consulted Edward on the making of those mechanical devices. Biweekly meetings of Franklin's newly founded Society of Political Inquiries were held there with gentlemen including David Rittenhouse and Francis Hopkinson, although Edward was never listed among those present.

Shortly before Franklin's death in 1790, Edward possibly was among the friends mentioned below by Dr. Jones, who heard Franklin describe a new final codicil to his will that reflected a lifelong admiration for artisans making things with their own hands, perhaps in part what drew him and Edward together. Franklin left substantial sums to the cities of Boston and Philadelphia to provide small loans to aspiring married craftsmen, recognizing that the financial help he received from two benefactors at the start of his printing career was crucial to his initial success.

Benjamin Franklin died Saturday night, April 17, 1790. His doctor John Jones (not Edward's son Benjamin) reported that during the statesman's final days he was "conversing chearfully with his family and a few friends, who visited him…"[376]

Franklin was not honored by an official state funeral, having recently antagonized many Federal officeholders by ardently petitioning for the abolition of slavery. But four days after his death, the funeral procession of dignitaries, eighty clergymen, and many APS members silently marched the short distance from his home in Franklin Court to the northeast corner of the Christ Church burial ground. Upwards of 20,000 mourners, more than half of the city's population, lined the route. The bells of every church tolled at 3 p.m. David Rittenhouse was one of six pallbearers, and Edward and his two grown sons possibly helped carry the casket or were in the group of family and close friends that followed closely behind it. Reverend William Smith (1727–1803) delivered the eulogy.

At the request of the APS, a second Franklin eulogy was pronounced by Reverend Smith on March 1, 1791, at the German Lutheran Church on Fourth Street. An APS vote was tied between

**Figure 9.6** Waywiser originally owned by Franklin, possibly the one bequeathed to Edward. Courtesy of Franklin Institute, 5259.

Smith and Rittenhouse for the honor, but Rittenhouse, a close friend of the departed, deferred to Smith, although Franklin had grown to detest that clergyman. President Washington, Vice-President Adams, their wives, the entire Cabinet and Supreme Court, the Senate, and House of Representatives endured the hour-long oration. Most likely Edward was there as well.

In Franklin's final will that was drafted during his eighty-fourth year, he wrote: "I request my friends, Henry Hill, Esq; John Jay, Esq; Francis Hopkinson, Esq; and Mr. Edward Duffield of Benfield, in Philadelphia county, to be the executors of this my last will and testament, and I hereby nominate and appoint them for that purpose."[377] The executor records at the APS indicate that Henry Hill, not Edward or Hopkinson, handled the bulk of the estate's finances and correspondence.

Three different men, including David Rittenhouse, were assigned to prepare the four-page estate inventory. It included a "Clock (on the stairs)" and another in the "Blue Room," each valued at £20. Neither clock is identified by maker and could have been made by Edward or John Whitehurst. A third unnamed timepiece in the library was bequeathed to grandson William Temple Franklin; it may be Catalogue No.4.

Another codicil added Franklin's specific bequest to Edward: "I request my friend Mr. Duffield, to accept moreover my French wayweiser, a piece of clock-work in brass, to be fixed to the wheel of any carriage." Rittenhouse valued it at £5 (Figure 9.6). A circa–1763 waywiser originally owned by Franklin is in the collection of the Franklin Institute[378] and may be Franklin's bequest. It advanced a mile indicator for every 400 rotations of a carriage wheel that was thirteen feet in circumference, enabling Franklin to record distances during his postal inspections. The pointers indicate 1,600 total miles traveled. Edward Duffield Neill wrote that his great-grandfather "also invented a horse-rake and an odometer, containing some improvements on a French instrument which Dr. Franklin left him by will."[379]

Franklin provided £60 to his executors for their duties. This may have been the source of £35 that Edward received the following year from Henry Hill "in full of Doctor Franklin's legacy to me."[380] Franklin's scientific instruments were bequeathed not to Edward but to Francis Hopkinson, "my ingenious friend."

Many of Franklin's possessions were dispersed soon afterwards by family members who were reducing clutter or raising money. Edward would have had an opportunity to acquire more mementos at an October 1790 auction of household and kitchen furniture that included a small clock belonging to Franklin's grandson: "At Mr. T. Franklin's Lodgings at Mr. Bache's, Market, between Third and Fourth-Streets… curious small time-piece, shewing the hour, moon's age, day of the week and month."[381]

Then in May, 1792, auctioneer Richard Footman advertised:

At the House of the late Dr. FRANKLIN, up Franklin Court, Market Street.… A variety of valuable Furniture and Plate, consisting of Mahogany side-Board, Dining, Card, and Pembroke Tables; Mahogany chairs; Looking-Glasses, Cloath's Presses; Tea Urns, Plated candlesticks, Windsor chairs, an elegant Sopha; chintz Window Curtains; Chests of Drawers; a Forte Piano; a Harpsicord; a Copying-Press.… Also A SEDAN CHAIR.…[382]

**Figure 9.7** Detail from *A Philosopher Lecturing on the Orrery* by Joseph Wright of Derby, 1766, possibly depicting John Whitehurst. Courtesy of Wikipedia.

One final note further affirms the close connections between Edward and Franklin. Edward is known to have had a Franklin portrait, not listed in his 1803 estate inventory, to remind him of his esteemed friend. It allegedly was painted by Benjamin West and was passed down to Franklin's great-great-grandson. The artist attribution was disputed in a 1906 article:

Edward Duffield, one of Franklin's executors, had a portrait was supposed to have been done by "West", but clearly could not have been unless he copied it. It now belongs to one of Franklin's descendants, Dr. Thomas Hewson Bache, of Philadelphia, and from its rigidity and hardness would seem without doubt to be a not very faithful copy of the portrait painted by Benjamin Wilson…[383]

Even if the Franklin portrait was a poor copy, Edward's enduring memories of his famous friend would have overcome its deficiencies. From Edward's earliest years until Franklin's death, the two men and their immediate families had close, personal relationships. While Franklin might conceivably have encouraged his younger friend to become a printer under his tutelage, we today are grateful that Edward chose horology instead. The existing documents are enough to confirm that, had we been flies on the walls of the rooms where the two men sat and conversed, we would have been even more certain of their mutual affection and the importance of their friendship throughout their long lives.

*APPENDIX I*

# EDWARD DUFFIELD'S WILL

Although the original of Edward's 1801 last will and testament seemingly has been viewed by past scholars, it cannot now be located by city employees. However, the author was able to view an official handwritten copy from the period and produced the following transcription.

"Be it Remembered that I Edward Duffield of Moorland Township in Philadelphia County, State of Pennsylvania (late of the City of Philadelphia clock and watchmaker) being in a moderate state of health and of a sound mind & perfect memory consider the uncertainty of life do make this my last will and testament in manner and form follows that is to say: In the first place I order and direct that all my just debts and funeral expenses be paid and satisfied. Item my old farm being the Northwestward plantation or part of my tract of land called Benfield situate in Moorland Township Philadelphia County & State of Pennsylvania bounded Southwestward by Lower Dublin township and land of Jon. Scholfield Esq. and John Swift and Northwestward also by the said John Swifts land Northeastward by Byberry Township and South Eastward by a line of stones set by marked trees which divides this from my middle plantation (part of Benfield) Beginning at a stone marked D.1771, set in the Public Road between Byberry & Moorland Township thence South fifty Degrees West along the said line of stones & marked trees across Moorland to a stone marked D.1778, set in the line of Lower Dublin Township. This farm I order & direct to be divided by a line beginning at the South Easternmost point of Jonathan Scholfields land where are two stones placed thence extending across Moorland to a middle point (in the line of Moorland and Byberry) between the Easternmost corner of John Swifts land and the Northernmost corner of the land hereinafter given to my daughter Sarah Hepburn, and a lane or road on the said farm sixteen feet six Inches wide by the line of Lower Dublin Township & continues on the land hereinafter given to my said daughter Sarah, extending from the South East most point of Jon. Scholfield's land where are two stones placed to the public road which crosses Moorland. This lane or road must be and remain for the use of common of the owners and occupiers of my three adjoining farms This farm or plantation being thus divided I give & devise the Northwesterly part thereof with the improvements & appurtenances unto my grandson Edward Duffield, son of my late son Benjamin Duffield dec. to hold the same to him the said Edward Duffield his heirs & assigns for ever with the right of the use of the said lane or road - and I give and devise the South Easterly part of the said farm or plantation with the improvements & appurtenances unto my grandson John Duffield son of my said son Benjamin deceased, and I also & devise unto my said grandson John Duffield a lot or piece of land adjoining situate in Lower Dublin Township which I bought of Stephen Decatur containing about ten acres with appurtenances to hold the same to him the said John Duffield his heirs & assigns for ever subject to the said lane or road. If either of my said grandsons Edward & John above named die under age and without lawful issue, I give and devise the land given him so dying with the appurtenances unto his surviving brother his three —- grand daughters) Catharine Martha & Rebecca to hold the same to him the sur.... other and his three sisters above named in equal parts or shares and to their ——- for ever as tenants in common and not as joint tenants. Item I give ———-(top of page 2 cut off) my middle plantation and tract of land and four tenements part of Benfield aforesaid, bounded Southwestward by Lower Dublin Township North Westward by the above described line of stone & marked trees which divides this from the aforesaid old farm or plantation North Eastward by Byberry Township & South Eastward by the aforesaid public road which crosses Moorland Township between this & the plantation whereon I now dwell and a lot or piece of land in Lower Dublin Township adjoining which I bought of Joseph Ashton containing about three acres with the improvements & appurtenances and all the stock and farming utensils on the said plantation which I provided for her, to hold the same to her my said daughter Sarah Hepburn her heirs and assigns for ever subject to the lane or road aforesaid. Item I give & devise unto my Daughter Elizabeth Ingraham all that tract of Land and Plantation situate partly in Byberry and partly in Moorland Townships, bounded by Potquessin Creek, by land formerly Nathaniel Waltons by land of Silas Walton, by a line of stones marked D.1792 which divides this from the place where I now dwell, by Lower Dublin Township and land of William Bell the same being composed of part of the land I purchased of Daniel Walton, the chief part fo the land and reversions which I purchased of James Ash Esq. Sheriff, a small lot which I bought of Silas Walton (all in Byberry) and part of Benfield in Moorland Township with the rent improvements and appurtenances and also all the tract of about twenty one acres of land more or less situate in Burlington County in the State of New Jersey, which I bought of the assignees of Cowperthwaite & wife with the rents improvements and appurtenances to hold the same to her the said Elizabeth Ingraham her heirs and assigns for ever accepting and reserving nevertheless unto Esther the wife of Peter Stats during her natural life only the use of the tenement and lot of ground near Potquessin where they now dwell. Item I give and devise unto my son Edward Duffield all that tract of land or plantation whereon I now dwell with the buildings improvements and appurtenances thereunto belonging bounded North Westward by the aforesaid public road which crosses Moorland between this and my middle plantation (part of Benfield) North Eastward by Byberry and land of Silas Walton until it comes to a stone marked D.1792, at the distance of fifty six perches from the Southernmost corner of Silas Waltons land, thence South fifty five degrees West to a stone set in the line between Byberry & Moorland Townships marked on the South west side (for Moorland) M.1792 on the North East side B. (for Byberry) and on the top 1792, thence the same course South fifty five degrees West across Moorland to a stone marked D.1792, set in the line of Lower Dublin Township; the same being part of the lands of Benfield in Moorland and part of the land I purchased of Daniel Walton in Byberry Township, together with all the farming utensils which I may have on this place at the time of my decease and my share of the Philadelphia Library, to hold the said tract of land or plantation as above described whereon I now dwell with the appurtenances and the farming utensils and my share of the said library, and my said son Edward Duffield his heirs and assigns for ever. Item I hereby direct and authorize my Executors herein after named or the survivor of them to sell my two dwelling houses and grounds thereto belonging situate in Philadelphia at the corner of Arch Street and Second Street also a lot of about five acres & three quarters of an acre situate in Moyamensing Township in Philadelphia County for the best price they can obtain for the same and to execute good and sufficient deed or deeds of conveyance therefor unto the purchaser or purchasers and his her or their heirs & assigns for ever - The money arising from these ———s I give in manner following to wit One fourth part thereof I give to my daughter Sarah Hepburn her heirs and assigns; on other fourth part thereof I give to my daughter Elizabeth Ingraham her heirs and assigns; one other fourth part thereof give to my son Edward Duffield his heirs and assigns; and the remaining fourth part I give to my three grand daughters Catharine Martha and Rebeca above named (daughters of my son Benjamin deceased their heirs & assigns in equal part or shares. Item I give and devise my two lots of ground containing together about four acres and an half having a tenement and barn there situate on the South West side of Hanover Street in Kensington (about three acres whereof I bought of Thomas Bond & one acre and a half or thereabout I bought of William Dewers Esq. Sheriff) and also a lot of ground twenty two feet on the North west side of Queen Street one hundred and eight feet deep near Hanover Street in Kensington (which I bought of Thomas Bond) in manner following to wit the said lots with their appurtenances being divided into four equal parts in value. One fourth part thereof I give to my daughter Sarah Hepburn her heirs and assigns; one fourth part more thereof I give to my daughter Elizabeth Ingraham her heirs and assigns; one fourth part more thereof I give to my son Edward Duffield his heirs & assigns; and the remaining fourth part I give to my three granddaughters Catharine Martha & Rebecca (daughters of my said son Benjamin deceased) and their heirs and assigns in equal parts or share as tenants in common. Item All the rest and residue of my estate real and personal not hereinbefore given devised or disposed of of whatever kind & wherever situate and being with all and every of the appurtenances I give & devise as follows to wit One fourth part of the said residue of my estate I give to my daughter Sarah Hepburn her heirs and assigns; One other fourth part thereof I give to my daughter Elizabeth Ingraham her heirs and assigns; one other fourth part thereof I give to my son Edward Duffield his heirs and assigns and the remaining fourth part I give to my late son Benjamin given children namely Edward, John, Catharine, Martha & Rebecca their heirs & assigns in equal parts or shares as tenants in common; and if any of them my said grand children die under age and without lawful issue the part of any of them so dying shall go to the survivors in equal parts or to the survivor if but one remains. Lastly I nominate constitute and appoint my son Edward Duffield and my son-in-law Francis Ingraham to be Executors of this my last will and testament on condition they render just accounts & pay all that shall appear justly due to my estate and hereby revoke all former wills by me made ratifying and confirming this and no other to be my last will and testament. In witness whereof I have hereunto set hand and seal the twenty eighth day of July in the year of our Lord one thousand eight hundred and one. My desire or request is to be buried by my late wife and that my executors provide a tombstone to cover both our graves having our names our ages and the times of our deaths engraved thereon the same to be charged with other funeral expenses. Edwd. Duffield (seal) Signed sealed published & declared by the above named Edward Duffield to be his last will and testament in the presence of us who hereunto subscribed our names as witnesses at the request of the testator and in his presence & in the presence of each other. "and pay" being first underlined in the 14th line of this page to be read instead of the erasement under it. Silas Walton, Thomas Powell, James Simpson Junr. NB The words (with the right of the of) interlined between the 24th & 25th lines from the top of the first page instead of the words (subject to) which I have erased are in my own hand writing and is intended to correct an error. Witness my hand March 6th 1802. Edwd. Duffield (seal). Philadelphia July 23rd 1803 there personally appeared Silas Walton, Thomas Powell, James Simpson Junr. the witnesses to the foregoing will, the said Silas and James on solemn affirmations and the said Thomas on oath did declare and say that they saw and heard Edward Duffield the testator sign seal publish and declare the same—for his last will and testament and that at the doing thereof he was of sound mind —- memory and understanding to the best of their knowledge & belief. Corain Chas. Swift—Edward Duffield one of the exor. sworn same day and Francis Ingraham —[final lines indistinct]."

APPENDIX II

# 1803 ESTATE INVENTORY OF EDWARD DUFFIELD

*Transcribed from hand-written copy made by the late Ian Quimby. Original estate inventory cannot be located by city employees. Amounts shown are in dollars.*

Winterthur Downs Library PH-77. "Inventory of Goods, Chattels, Rights and Credits belonging to the the Estate of Edward Duffield deceased viz:

| | |
|---|---|
| Amot. Principal & Interest on bonds etc. until 12th July 1802 | 11936.25 |
| Mathew Clarkson bond & Interest | 525.46 |
| Ditto | 193.75 |
| Shares in Bank Stock etc. | 4788.57 |
| Cash in Bank MS | 536.52 |
| in Drawer | 198.69 |
| Recd for Grain | 88.92 |
| Recd for Iron | 14.13 |
| Outstanding debts | 594.72 |
| Balance of Philip Ecker's Acct | 37.49 |
| of Richard Duffields Acct | 47.54 |
| **Household Furniture etc.** | |
| 6 Walnut Chairs | 7.25 |
| 1 Arm. Do | 1 |
| 1 Looking Glass Walnut Frame | 10 |
| 1 Dining Table | 2 |
| 1 Small Do | 1 |
| 1 Walnut Tea Chest | 3 |
| 1 Small Carpet | 3 |
| 1 Pr Brass Andirons Shovel & Tongs | 7 |
| 1 Square Japan'd Waiter | 1 |
| 1 Picture | .50 |
| 11 Knives & Forks Ivory Handles | - |
| 12　　do　　Desert | 10 |
| 12　　do. Ivory. do | - |
| 8 Pots Extract Butter Nut | 4 |
| 1 Basket & Tumblers, 1 Wine Glass 16 Jelly | |
| & 4 beer, in long Basket | 1.50 |
| 1 Basket containing 2 Slats, 11 Wine Glasses, | |
| & 4 Tumblers | 1.50 |

| | |
|---|---|
| 5 Knives & 6 Forks. Real Buck | 2 |
| 10 Coffee Cups & Saucers Bowl Tea Caddy | 2 |
| 12 Tea Cups Coffee Pot, Bowl, Tea Caddy 2 qt. & sugar dish | - |
| 3 Wooden round boxes | .37 |
| 1 Mahogany Desk | 30 |
| 1   do   Dining Table | 10 |
| 1 Walnut Breakfast Table | 3 |
| 1 Sopha | 6 |
| 1 Time Piece Walnut Case | 45 |
| 1 Small case cont Instruments | 3 |
| 1 Shagreen case Watch | 8 |
| 6 Windsor Chairs | 1 |
| 1 arm chair | 1 |
| 6 prints | 3 |
| 1 walnut candle stand | 1 |
| 1 looking glass mahogany frame | 10 |
| 1 chimney     do | 8 |
| 1 umbrella (silk) | 2 |
| 1 pair brass andirons. Shovel & tongs | 6 |
| 4 Japan'd waiters | 1 |
| 1 pair pistols | 3 |
| Birch's Views | 3 |
| A small basket containing 10 nut picks | 1 |
| A perambulator | 2 |
| 7 decanters | 2 |
| Casters walnut frame | 1.50 |
| 2 Coasters | .25 |
| China bowl, butter tub, tea pot, 3 san - 2 cups, cocoanut sugar dish | 3 |
| Gallon bottle, a pearl shell | .25 |
| Old basket | .12 |
| Walnut shaving case | 2 |
| Clothes brush, red cedar faucet sundry | 2.25 |
| 1 pine table. Stair carpet | .50 |
| 12 custard cups, marble mortar | 3 |
| 4 brass candle sticks, stone jug | 4 |
| 2 Japan'd. Do. Walnut tray | .75 |
| 3 dishes queens ware, oil cloth umbrella | 2 |
| Set of surveying instruments | 8 |
| Case containing gin | 3.50 |
| 1 lanthorn | 1 |
| 1 pair lime squeezers, stone jug | 2 |
| Round mahogany waiter & sundry glass | .25 |
| Tin scales & weights | .50 |
| 2 large canisters, cedar firkin (?) | 2.50 |
| 1 mahogany dressing table | 4.50 |
| 1 looking glass | 6 |

| | |
|---|---|
| 6 chairs with covers @1.50 | 9 |
| Brass andirons, shovel & tongs | 4 |
| 1 easy chair | 5 |
| 1 bed stead, stained, bed curtains | 43 |
| 2 window curtains | 2 |
| 1 trundle bed stead | 2 |
| 36 plates, 17 dishes, a tureen, 2 sauce boat | 12 |
| Brass shovel 2 tongs | 2 |
| 5 blankets 2 gowns | 12 |
| 1 bed stead. Bed & pillow | 36 |
| 4 chairs | 4 |
| 2 arm do | 1.50 |
| 1 mahogany chest of drawers | 16 |
| 1 walnut bureau | 5 |
| 1 dressing glass 1 walnut trunk | 4 |
| 1 square chest | 1 |
| 1 round tea table mahogany | 2.50 |
| 1 double barrel gun, powder flask & shot bag | 20 |
| 1 single    do | 6 |
| 1 pair steel andirons | 2 |
| 1 small trunk, spinning wheel | 1 |
| 14 pair sheets | 40 |
| 13 table cloths | 35 |
| 4 napkins | 2 |
| 6 pair pillow cases | 3 |
| 5 counterpanes | 30 |
| 1 bed stead, bed, bolster, pillow & bed clothes | 30 |
| 1 looking glass | 1 |
| 1 walnut bureau | 2.50 |
| 1 mahogany dressing table | 1.50 |
| 1 looking glass walnut frame | 4 |
| 4 walnut chairs | 2 |
| 4 arm.  Do | 2 |
| 1 bed stead, bed etc. | 30 |
| 1 looking glass, clothes horse, & ironing board | 1.50 |
| 4 blankerts | 4 |
| Andirons shovel & tongs | 3 |
| Tin measure & funnel | .50 |
| 2 hatchets | 3 |
| 3 servants bedsteads, bed and bedding | 36 |
| 1    do         do | 8 |
| 1 arm chair. 1 pair coarse sheets | 2 |
| 1 old saddle & saddle bags | 2 |
| 2 cases with liquor. 1@10 & 1 @ 30 | 40 |
| 2 copper tea kettles. 2 plate baskets | 2 |
| 10 jars extract butter nut | 4 |
| 1 pair brass sconces | .25 |

| | |
|---|---|
| 1 round tea table | 1 |
| 1 old chest drawers | .25 |
| 1 prospect city of Philadelphia | .50 |
| 1 bed stead bed & bedding | 10 |
| 1 arm chair | 1 |
| 4 blankets. 4 coverlets | 12 |
| 1 bedstead & bed | 10 |
| 2 arm chairs | 2 |
| Ticking | 2 |
| Map of Pennsylvania | .50 |
| 2 small wheels & large reel | 4 |
| 2 large chests | 4 |
| 3 firkins | 1 |
| 23 pewter plates | 6 |
| 5 dishes | 5 |
| 5 water plates | 3 |
| 7 chairs | 1.75 |
| 1 table (walnut) | .75 |
| 1 dough tray | 1 |
| 7 sad irons | 1.75 |
| Shovel, tongs & andirons | 4 |
| 1 jack. 1 spit | 4 |
| 1 pair iron candlesticks | .50 |
| Brass kettle | 6 |
| 4 iron pots | 6 |
| 1 iron kettle | 3 |
| 1 Dutch oven & bake iron | 1.50 |
| Colander | .25 |
| Chopping tub & knife | 2 |
| 3 tubs 2 baskets | 2 |
| 1 lamp | .25 |
| 2 pewter basins & salt cellar | .75 |
| 2 stone jugs, coffee pot & Skillet | 1 |
| 2 flour buckets | .50 |
| 1 open stove | 6 |
| 1 ten plate stove | 8 |
| 7    do    @16 | 112 |
| 2 firkins | 2 |
| 2 meat tubs | 6 |
| 25 iron bound barrels | 25 |
| Stone jar | .25 |
| Soap tub & 2 fish barrels | .75 |
| Candle tub | .67 |
| 2 empty demijohns | 3 |
| 4 demijohns with whiskey | 20 |
| 1 watch case | 15 |
| 17 bottles wine | 10 |
| Plate 14 lbs 11 oz 20 @8/0 oz | 229.24 |

| | |
|---|---:|
| Gold 56 dwt 68 @6/p dwt | 453 |
| 1 silver mounted turtle shell power flask | 6 |
| 1 turtle shell bread or cake basket | 6 |
| 2 large tubs | 16 |
| 4 pump logs | 15 |
| Sleigh | 12 |
| Large turning wheel | 2 |
| Old chair 2 wheels | 10 |
| 1 cart hook 2 cross bars | 2.25 |
| 4 pointed shovels | 4 |
| 2 picks & 1 well cleaner | 1.50 |
| Tackle & falls, 1 beef rope | 5 |
| Land riddle | .50 |
| 11 screw augers different sizes | 4 |
| 6 pod augers | 1.50 |
| 1 long auger | 1.50 |
| 1 hold fast, 1 iron driver | 1 |
| 3 wooden mallets, 2 hammers | 1 |
| 1 broad axe 1 square | 1.25 |
| 2 hand saws 3 drawing knives | 3 |
| 2 tenant. do 1 bevel | 3 |
| 2 compass. do 1 spoke shave | 1 |
| 2 jack planes 1 brace bit | 2 |
| 2 double iron fore planes | 1.50 |
| 2 single do plough & grove | 2 |
| 1 pointer | .75 |
| 2 smoothing planes 1 other plane | 1 |
| 57 planes various sizes @ 125/10 | 14.25 |
| 2 hollow hounds | 1 |
| 3 bound boxes winch & well rope | 2 |
| 4 spoke gimlets, 3 smaller etc. | .75 |
| Plough plane | 1 |
| 2 large gouges | 1 |
| 33 chisels different sizes | 4.50 |
| 2 small gouges | 1 |
| 1 " brass square & compasses iron | .75 |
| 1 screw driver | .33 |
| 4 old files | .25 |
| 1 taper bit | .12 |
| 1 pair calipers 1 large pair iron compasses. | 1.50 |
| 1 oil stone & box | 1 |
| 1 plumb rule & setting knife | 6 |
| 1 coopers ? | 1 |
| 1 bell, 1 small level, 1 small grindstone | .75 |
| 1 sash holder | .15 |
| 1 old pair bellows | 2 |
| 2 anvils | 2 |
| 1 peak iron | 2 |

| | |
|---|---|
| 8 hammers | 2 |
| 2 sledges | 2 |
| 2 mason hammers | .75 |
| 6 pair smiths tongs | 1.50 |
| 2 winches | .50 |
| 2 old files | .12 |
| 1 brass rule | .25 |
| 1 set blowing tools | 1 |
| 3 screw plates & taps | 3 |
| 3 stone cutters chisels | .75 |
| 1 set pump auguers & shank | 15 |
| 1 plumbers tools | 1 |
| Sundry old iron | 10 |
| 2 carriage boxes | .50 |
| 2 meal chests | 6 |
| 1 rafter level | .50 |
| 5 empty casks different sizes | 1 |
| 2 brass hoops | .25 |
| A quantity of rye | 171.69 |
| A quantity of buckwheat | 106.15 |
| A quantity of damaged do | 22 |
| A quantity of oats | 220 |
| A quantity of hay | 16 |
| 2 vinegar casks | 2 |
| 2 empty oil barrels | .50 |
| A jar with oil | 34 |
| 2 empty hgd | .50 |
| 1 hgd | .50 |
| 5 cross hoops | 1 |
| 2 spouts | .75 |
| 2 empty casks | .50 |
| A quantity of corn 649 bush 58 lbs @ 4/ | 403 |
| 5 walnut screws | 20 |
| 8 poplar planks | 2.40 |
| 10 in. Boards | 2 |
| 4 empty casks | 2 |
| 3 axel trees | 1.10 |
| 7 … pig iron | 1 |
| 1 beams | 24 |
| 1250 ft oak plank | 20 |
| 6 empty hhds | 3 |
| 7 tierces | 6 |
| 10 barrels | 10 |
| 4 casks | 1 |
| 3 long ladders 1 short | 2.50 |
| 1 box scales & weights | 1 |
| A quantity of wheat in straw - | |
|     do    do at John Terry's Farm | 146.93 |

| | |
|---|---|
| do      barley in straw | 44.69 |
| Lye cask | 2 |
| 3000 ft white pine boards @ 1/? | 45 |
| 1041 ft sap boards @`0/ | 13.87 |
| 1260 ft lath @6/ | 10.08 |
| 2300 shingles 16.50/100 | 41.25 |
| 2440 staves @5 | 32.44 |
| Loose scantling & lath | 2.50 |
| 343 perch stone at quarries 3/ | 137.20 |
| 300 bricks. 7/6 | 3 |
| 2000… rails. 5/6.25/100 | 125 |
| 1 mill shaft | 4 |
| Cut stone | .50 |
| 100 cogs @3 | 3 |
| Oak wood 23 1/4 cords @3.50/1 | 81.37 |
| 1 broad axe | 1.50 |
| 1 fowling piece | 6 |
| 1 double iron & 4 single iron smoothing planes | 1.33 |
| 1 tooling(?) dish 4 shingling hatchets | 1 |
| 15 plane bits | 1.50 |
| 1 box containing glass | 5 |
| 4 kegs | 2 |
| Sundry fishing tackle | 3 |
| Gun stocks | 4 |
| A quantity of sheet lead | 21.75 |
| 70 lbs nails @10 | 7 |
| Shovel tongs andirons & 2 fender | 3 |
| 1 saw, brass mortar & iron pestle | 1.50 |
| 1 drawer containing sundries | 2 |
| 1 do with screws & do | 4 |
| 1 turkey stone | 1 |
| 30 bottles sherry wine | 9 |
| 2 trowels & 1 paving tool | 1 |
| 10 chisels 2 butcher knives, 1/2 in. Auger. | 1.75 |
| 2 padlocks, 2 paint brushes | .75 |
| Sundry corks | .30 |
| 3 con bells | 1 |
| A quantity of real leather | 1 |
| 191 gall. Beer | 24 |
| Books | 85 |
| Sundry chisels & punches | 1 |
| 1 rivet mould, 2 nail do | .75 |
| 218 ft oak plank | 3.47 |
| 1 batteau & oars | 13.50 |
| Balance of David Clarks acct. omitted | 5.99 |
| 14 1/2 lb whale teeth | 6.44 |
| 4 lb 11oz turtle shell | 14.6 |
| Lot of seals & old springs | 12 |

| | |
|---|---:|
| Lot of files | 20 |
| 4 doz. 2/12 do | 4.12 |
| Lot brass wire plate etc. | 13 |
| Lot | 5.29 |
| 50 lbs sheet brass @2/4+1 piece 25/100 | 15.81 |
| 143 1/2 old do @14d | 22.91 |
| 3 1/4 lbs gun powder | 1.75 |
| 11 3/4 lbs shot | 1.50 |
| 3 oz 2dwt silver bows and studs | 3.10 |
| 1 brass flask | 2 |
| 1 small & 1 open stove | .46 |
| Sundry files | 6.26 |
| Hand vises | 1 |
| Hammers & tools | .37 |
| Sundry tools | .63 |
| Screw plate & tools | 1.20 |
| Gun lock & mountings strike? 1 compass. | 1.97 |
| Oil stone saw chain | 2.80 |
| 2 boxes with drawers & lot of steel | 1.50 |
| 1 desk with drawers | 1.47 |
| Ivory watch tools & vises | 2.25 |
| Seal files & punches | 4 |
| Brass steel turtle shell etc | 1.92 |
| 2 clocks compleat @$16 | 32 |
| 3 do unfinished | 42 |
| 1 do musical do | 8 |
| 1 alarm | 10 |
| 1 small anvil & 6 small tools | 1.97 |
| 2 bench vises | 2 |
| Lot of ivory, wire etc. | ? |
| 1 do files & sundry tools | 1.97 |
| 1 lot watch crystals, materials, lamp? | 10.90 |
| 14 demijohns with whiskey omitted in page | 5.56 |
| 2 gallon bottles 1 flask & 1 funnel.   Do | 1 |

We the subscribers do certify the [unclear] Inventory was appraised by us on the twenty-ninth day of September 1803

    Joshua Comly

    Derick Peterson

    Jacob Sommer

D. Peterson one of the appraisers sworn the 17th day of Sept. 1804 before J. [unclear] Rep.Reg.

# APPENDIX III

# EDWARD DENT'S 1837 LIST OF 992 PARTS IN A POCKET WATCH AND THE FORTY-THREE ARTIFICERS EMPLOYED TO CREATE THEM

The 1837 *Journal of the Franklin Institute of the State of Pennsylvania and Mechanic's Register*....* recorded on pages 139–140 that renowned British horologist Edward Dent, of the firm Arnold and Dent, computed and demonstrated all the labor and materials required at that time to produce a pocket watch. A half-century earlier, during Edward Duffield's lifetime and when nearly all watches had verge escapements, the situation was similar. Only a very few watchmakers, working at the highest levels of the craft, produced watches entirely by themselves. Even if the 826 pieces of a fusee chain are deducted, this still leaves 166 tiny parts to be expertly fabricated. Following is my transcription of the Journal's report:

Watch Statistics

Mr. Dent (Arnold and Dent), in his illustrations at a lecture on the construction of watches and chronometers, given by him at the royal Institution on the 7th ult., laid before the meeting the dissection of a detached lever watch (compensation-balance,) every part was separated and displayed, but grouped in one of six larger divisions to which it belonged.

Each part had been previously examined, and its distinct constituent pieces counted by the lecturer; the surprising result of this enumeration was exhibited in a table, of which we lay a copy before our readers. In addition, will be found the number of kinds of artificers concerned in the operations necessary for the construction of a good watch. When to these are added the amount of previous operations which the materials constituting each piece must undergo, *before* it comes into the hand of the watch-artificer, a glimpse may be obtained of the extensive and numerous changes of form and value which "raw material" receives in its progress, from the mine to so refined a manufacture as a finished watch.

| Parts | No. Of Pieces | Trades Employed |
|---|---|---|
| Pillars | 4 | 1 |
| Frame | 4 | 1 |
| Cock and Potence | 2 | 1 |
| Barrel and Arbor | 3 | 1 |
| Going-Fuzee | 14 | 2 |
| Wheels | 4 | 1 |
| Pinions | 4 | 2 |
| Stop-Stud | 1 | 1 |
| Stop and Spring | 3 | 1 |

| Parts | No. Of Pieces | Trades Employed |
|---|---|---|
| Click and Ratchet | 3 | 1 |
| Motion | 16 | 2 |
| Jewels (5 holes) | 28 | 2 |
| Cap | 3 | 2 |
| Dial | 5 | 3 |
| Index | 1 | 1 |
| Escapement | 13 | 3 |
| Compensation Balance | 9 | 1 |
| Case | 3 | 1 |
| Pendant | 2 | 1 |
| Case-Joint | 6 | 1 |
| Case-Spring, &c. | 4 | 2 |
| Main-Spring | 1 | 2 |
| Chain | 826 | 3 |
| Hands | 3 | 1 |
| Glass | <u>1</u> | 1 |
| | | |
| Total of Pieces | 992 | |
| Engine Turner | 1 | |
| | | |
| Engraver | 1 | |
| Gilder | 1 | |
| Examiner | <u>1</u> | |
| | | |
| Total of kinds of Artificers employed. | 43 | |

\* I am grateful to the Boston Athenaeum, of which I am a Proprietor. Their shelves contain many volumes of this journal and countless other old and obscure printed publications that are extremely difficult to find and study elsewhere.

# JOHN WOOD,
## WATCH AND CLOCKMAKER,
### In Front (the corner of Chesnut) street,
#### HAS FOR SALE,

Watches,
Clocks,
Clock-movements,
Finished Faces,
Bells, Hands,
Catgut,
Slit Pinions,
Clock and Watch-makers' Turn-benches,
Hand Vices,
Plyers,
Cutting Nippers,
Sliding Tongs,
Tail Vices,
Tweefers,
Screw-keys,
Cafe Stakes and Hammers,
Turkey Oilstones,
Screw Dividers,

Blow Pipes,
Calibres,
Screw-drivers,
Silversmiths Hand Shears,
Watch Main-springs,
Fufee Chains,
Glasses,
Gold, Gilt, and Steel Watch-bands,
Gold, filver, and steel Pendents,
Ditto cafe buttons,
Cafe Springs,
Pinion Wire,
Click ditto,
Gilding Scratch Brushes,
Adjusting Tools,

Spring Blowers,
Enamelled Dial Plates,
Steel Balances,
Endless Screws,
Verges, Pinions,
Motion Work,
Broaches,
Gravers,
Small Hammers,
Pendulum Wire,
Magnifying Glasses,
Clock and Watch Files,
Watch Keys,
Gilt and Steel Chains,
Silk Strings,
Seals, &c. &c.

Advertisement by Philadelphia watchmaker and merchant John Wood, Jr., selling horological tools and parts, *The Pennsylvania Evening Herald and the American Monitor,* September 8, 1785, p.56. *Courtesy of GenealogyBank.*

# BIBLIOGRAPHY

More than 250 books and articles, dating back to 1673, were read, studied, and referenced in preparing this comprehensive biography. Most are in the author's personal library. The bibliography is divided into sections on Philadelphia's colonial history, its furniture and decorative arts, and horological materials.

## PHILADELPHIA'S COLONIAL HISTORY

### BOOKS

Arbour, Keith. *Benjamin Franklin's First Government Printing: The Pennsylvania General Loan Office Mortgage Register of 1729, and Subsequent Franklin Mortgage Registers and Bonds.* Philadelphia: American Philosophical Society, 1999.

Baltzell, E. Digby. *Puritan Boston and Quaker Philadelphia: Two Protestant Ethics and the Spirit of Class Authority and Leadership.* New York: The Free Press, 1979.

Bell, Whitfield J., Jr. *Towards a National Spirit: Collecting and Publishing in the Early Republic to 1830.* Boston: Trustees of the Public Library of the City of Boston, 1979.

_____ and Smith, Murphy D. *Guide to the Archives and Manuscript Collections of the American Philosophical Society.* Philadelphia: The American Philosophical Society, 1966.

_____ and Greifenstein, Charles B. *Patriot Improvers: Biographical Sketches of Members of the American Philosophical Society, Volume Three, 1767-1768.* Philadelphia: The American Philosophical Society, 2010.

Betts, Charles Wyllys. *American Colonial History Illustrated by Contemporary Medals.* New York: Scott Stamp and Coin Company, 1894.

Bordley, J.B. *Essays and Notes on Husbandry and Rural Affairs.* Philadelphia: Printed by Budd and Bartram, For Thomas Dobson, 1799.

Boudinot, J.J. (ed.). *The Life: Public Services, Addresses and Letters of Elias Boudinot, LL.D., President of the Continental Congress.* Boston: Houghton, Mifflin and Company, 1896.

Bridenbaugh, Carl. *Cities in Revolt: Urban Life in America 1743–1776.* New York: Oxford University Press, 1955.

Bridenbaugh, Carl and Jessica. *Rebels and Gentlemen: Philadelphia in the Age of Franklin.* New York: Reynal & Hitchcock, 1942.

Butterfield, L.H. (ed.). *Letters of Benjamin Rush.* Published for The American Philosophical Society by Princeton University Press, 1951

Carey, Mathew. *A Short Account of the Malignant Fever Lately Prevalent in Philadelphia…* Printed by the Author, January 16, 1794.

Chaplin, Joyce E. *The First Scientific American: Benjamin Franklin and The Pursuit of Genius*. New York: Basic Books, 2006.

Clark, Edward L. *Record of the Inscriptions on the Tablets and Grave-Stones in the Burial-Grounds of Christ Church, Philadelphia*. Philadelphia: Colins, Printer, 1864.

Cohen, I. Bernard. *Benjamin Franklin's Science*. Cambridge, MA: Harvard University Press, 1990.

[Corlette, Suzanne]. *The Pulse of the People: New Jersey 1763–1789*. Trenton: New Jersey State Museum, 1976.

Crane, Elaine Forman (ed.). *The Diary of Elizabeth Drinker*. Boston: Northeastern University Press, 1991.

Davies, Benjamin. *Some Account of the city of Philadelphia, the capital of Pennsylvania, and seat of the Federal Congress; of its civil and religious institutions, population, trade, and government*. Printed by Richard Tolwell, 1794.

Dolin, Eric Jay. *Rebels at Sea: Privateering in the American Revolution*. New York: Liveright (Norton), 2022.

Dorr, Rev. Benjamin, D.D. *A Historical Account of Christ Church, Philadelphia, From Its Foundation, A.D. 1695, to A.D. 1841: and of St. Peter's and St. Jame's, Until the Separate of the Churches*. Philadelphia: R.S.H. George, 1841.

Drinker, Cecil K. *Not So Long Ago: A Chronicle of Medicine and Doctors in Colonial Philadelphia*. New York: Oxford University Press, 1937.

Duane, William (ed.). *Extracts from the Diary of Christopher Marshall Kept in Philadelphia and Lancaster During the American Revolution, 1774–1781*. Albany: Joel Munsell, 1877.

Duché, Jacob (as Tamoc Caspipina). *Observations on a Variety of Subjects, Literary, Moral and Religious, In a Series of Original Letters*. Philadelphia: Printed by Robert Bell, 1774.

Duffield, Edward. "Answers to Queries on Plaister of Paris by Mr. Edward Duffield, of Lower Dublin Township, Philadelphia County." *Agricultural Enquires on Plaister of Paris. Also, Facts, Observations and Conjectures on That Substance When Applied as Manure*. Philadelphia: Printed by Charles Cist, 1797.

Eberlein, Harold Donaldson. *The Architecture of Colonial America*. Boston: Little, Brown, and Company, 1915.

Eberlein, Harold Donaldson, and Lippincott, Horace Mather. *The Colonial Homes of Philadelphia And Its Neighborhood*. Philadelphia: J.B. Lippincott Company, 1912.

Eisenhart, Luther P. (ed.) *Historic Philadelphia From the Founding Until the Early Nineteenth Century: Papers Dealing with its People and Buildings with an Illustrative Map*. Philadelphia: American Philosophical Society, 1953.

Faris, John T. *Old Roads Out of Philadelphia*. Philadelphia: J.B. Lippincott Company, 1917.

Fenton, Will (ed.). *Ghost River: The Fall and Rise of the Conestoga*. Philadelphia: Library Company of Philadelphia, 2019.

Fortune, Brandon Brame, with Warner, Deborah J. *Franklin & His Friends: Portraying the Man of Science in Eighteenth-Century America*. Washington, D.C.: Smithsonian National Portrait Gallery in association with the University of Pennsylvania Press, Philadelphia, 1999.

Franklin, Benjamin. *A Catalogue of Books Belonging to the Library Company of Philadelphia*. Philadelphia: Printed by B. Franklin, 1741.

_____ *Some Account of the Pennsylvania Hospital; From its first Rise, to the Beginning of the Fifth Month, Called May, 1754*, Printed by B. Franklin, and D. Hall, 1754.

_____ *The Complete Poor Richard Almanacks* Reproduced in Facsimile with an Introduction by Whitfield Jr.. Bell, Jr., Volume II 1748–1758. Barre, MA: Imprint Society, 1970.

Fried, Stephen. *Rush: Revolution, Madness, and The Visionary Doctor Who Became a Founding Father*. New York: Crown, 2018.

Garvan, Beatrice B. *Federal Philadelphia 1785-1825: The Athens of the Western World*. Philadelphia: Philadelphia Museum of Art, 1987.

Graham, Daniel A. *The Family of John Potts*. Ellicott City, MD: privately printed and distributed, December 2006.

_____. *John Potts Jr., 1738–c1800 of Pottstown, Pennsylvania and His Descendants Through Grandchildren; The Loyalist*. Pottstown, PA: Pottsgrove Manor Historical Site, 2020.

Hamilton, Alexander. *Hamilton's Itinerarium: Being a narrative of a journey from Annapolis, Maryland, through Delaware, Pennsylvania, New York New Jersey, Connecticut, Rhode Island, Massachusetts and New Hampshire, from May to September, 1744*. Printed by Createspace, North Charleston, SC.

Hardie, James. *The Philadelphia Directory and Register*. Philadelphia: 1793.

Hastings, George Everett. *The Life and Works of Francis Hopkinson*. Chicago: The University of Chicago Press, 1926.

Hazard, Willis P. and Watson, John F. *Annals of Philadelphia, and Pennsylvania, in the Olden Times; Being a Collection…..* Philadelphia: Edwin S. Stuart, 1887.

Hewes, Lauren B. and Wolverton, Nan (eds.) *Beyond Midnight: Paul Revere* (to accompany an exhibit September 2019-October 2020). Worcester, MA: American Antiquarian Society, 2019.

Historical Society of Pennsylvania. *Guide to the Manuscript Collections of the Historical Society of Pennsylvania*, Philadelphia: 1991.

Hogan, Edmund. *Prospect of Philadelphia and Check on the Next Directory, Part I*. Philadelphia: Printed by Francis & Robert Bailey, 1795.

Horle, Craig W. et al. *Lawmaking and Legislators in Pennsylvania: A Biographical Dictionary*. Volume Two, 1710-1756. Philadelphia: University of Pennsylvania Press, 1997.

Jackson, Joseph. *America's Most Historic Highway: Market Street, Philadelphia*. Philadelphia: John Wanamaker, 1926.

_____ *Iconography of Philadelphia.* Privately printed by H.H. Kynett for the Membership of The Poor Richard Club, 1934.

Kalm, Peter. *Travels Into North America.* Barre, MA: The Imprint Society, 1972.

Keagle, Matthew. "Bespoke a suit of Regimentals: Clothing, Class and Place in Revolutionary Philadelphia, ARC Writing Sample.doc.

Keane, John. *Tom Paine: A Political Life.* Boston: Little, Brown and Company, 1995.

Kelley, Joseph J., Jr. *Life and Times in Colonial Philadelphia.* Harrisburg, PA: Stackpole Books, 1973.

Kidd, Thomas S. *Benjamin Franklin: The Religious Life of a Founding Father.* New Haven: Yale University Press, 2017.

Labaree, Leonard W. *The Papers of Benjamin Franklin.* New Haven: Yale University Press, 1959.

Lancaster, Bruce (narrative by). *American Heritage History of the American Revolution.* New York: ibooks, 1971 and 2003.

Lemay, J.S. Leo. *Ebenezer Kinnersley: Franklin's Friend.* Philadelphia: University of Pennsylvania Press, 1964.

_____ *The Life of Benjamin Franklin Volume 3: Soldier, Scientist, and Politician, 1748-1757.* Philadelphia: University of Pennsylvania Press, 2009.

Library Company of Philadelphia. *A Catalogue of Books Belonging to The Library Company of Philadelphia.* Philadelphia: 1789. (Edward Duffield listed among members.)

Looney, Robert F. *Old Philadelphia in Early Photographs 1839-1914.* New York: Dover Publications, Inc., Published in Cooperation with The Free Public Library of Philadelphia, 1976.

Lopez, Claude-Anne and Herbert, Eugenia W. *The Private Franklin: The Man and His Family.* New York: W.W. Norton & Company, Inc., 1975.

Lyons, Jonathan. *The Society for Useful Knowledge: How Benjamin Franklin and Friends Brought the Enlightenment of America.* New York: Bloomsbury Press, 2013.

Martindale, Joseph C., M.D. *The Gilbert Family, The Carver Family and The Duffield Family.* Frankford, PA: Martin & Allardyce, 1911.

_____ and Dudley, Albert W. *A History of the Townships of Byberry and Moreland, In Philadelphia, Pa.,* Philadelphia: George W. Martin, 1901.

McCulloch, J. *McCulloch's Pocket Almanac For the Year 1801.* Philadelphia: J McCulloch, 1801.

Mease, James M.D. First American Edition, with Additions, to Willich, A.F.M. M.D. *Domestic Encyclopedia: or, A Dictionary of Facts, and Useful Knowledge....* Five Volumes. Philadelphia: William Young Birch, and Abraham Small, 1803.

Menkevich, Joseph J. *3322 Willits Road - The Lower Dublin Academy - Proposal for Review & Consideration of Historic Designation By the Philadelphia Historical Commission.* 2016.

Meyer, Michael. *Benjamin Franklin's Last Bet.* Boston/New York: Mariner Books, 2022.

Miller, Lillian B. (ed.). *The Selected Papers of Charles Wilson Peale and His Family*, Volume 1. Published for the National Portrait Gallery, Smithsonian Institution, by Yale University Press, 1983.

Milley, John C. *Treasures of Independence: Independence National Historical Park and Its Collections.* New York: Mayflower Books, 1980.

Milnor, William. *A History of the Schuylkill Fishing Company of the State of Schuylkill,* Philadelphia: Members of the State in Schuylkill, 1889.

Mittelberger, Gotltlieb. *Journey to Pennsylvania.* Edited and Translated by Oscar Handlin and John Clive. Cambridge, MA: The Belknap Press of Harvard University Press, 1960.

Moss. Roger W. *Historic Houses of Philadelphia.* Philadelphia: University of Pennsylvania Press, 1998.

Nash, Gary B. *The Urban Crucible: The Northern Seaports and the Origins of the American Revolution.* Cambridge, MA: Harvard University Press, 1979.

Neill, Edward Duffield. *John Neill of Lewes, Delaware, 1739, and His Descendants.* Philadelphia: privately printed, 1875.

_____ *Biographical Sketch of Doctor Jonathan Potts,…..* Albany: J. Munsell, 1863.

Padelford, Philip (ed.). *Colonial Panorama 1775: Dr. Robert Honyman's Journal for March and April.* San Marino, CA: The Huntington Library, 1939.

Parrish, Samuel. *Some Chapters in the History of the Friendly Association For Regaining and Preserving Peace with the Indians by Pacific Measures.* Philadelphia: Friends' Historical Association of Philadelphia, 1877.

Parson, Jacob Cox. *Extracts from the Diary of Jacob Hiltzheimer, of Philadelphia.* Philadelphia: Press of Wm. F. Fell & Co., 1893.

Pennsylvania Archives. *Proprietary Tax List of Philadelphia County & City 1769.* Westminster, PA: Family Line Publications, 1988.

Peters, Richard. *Agricultural Enquiries on Plaister of Paris. Also, Facts, Observations and Conjectures on That Substance When Applied as Manure.* Philadelphia: 1797.

Powell, J.H. *Bring Out Your Dead: The Great Plague of Yellow Fever in Philadelphia in 1793.* Mansfield Centre, CT: Martino Publishing, 2016.

Rappleye, Charles. *Robert Morris: Financier of the American Revolution.* New York: Simon & Schuster, 2010.

Reutlinger, Dagmar E. *The Colonial Era in America.* Worcester, MA: Worcester Art Museum, 1975.

Richardson, Edgar P. et al. *Charles Willson Peale and His World.* New York: Harry N. Abrams, Inc., 1982.

Rush, Cary H. *All Saints' Church: A History of its Birth and Growth to Maturity 1772 to 1916.* Philadelphia: All Saints' Church, 1975.

Salinger, Sharon V. *"To Serve well and faithfully": Labor and indentured servants in Pennsylvania, 1682-1800.* Cambridge, UK: Cambridge University Press, 1987.

Sargent, Winthrop (ed.). *The Loyal Verses of Joseph Stansbury and Doctor Jonathan Odell Relating to the American Revolution.* Albany, NY: J. Munsell, 1860.

Scharf, J. Thomas, and Westcott, Thompson. *History of Philadelphia 1609-1884.* Philadelphia: L.H. Everts, 1884.

Schlesinger, Arthur Meier. *The Colonial Merchants and The American Revolution, 1763-1776.* New York: Columbia University, 1918.

Sellers, Charles Coleman. *Charles Willson Peale.* New York: Charles Scribner's Sons, 1969.

Seymour, Joseph. *The Pennsylvania Associators, 1747–1777.* Yardley, PA: Westholm, 2012.

Smith, Billy G. *The "Lower Sort": Philadelphia's Laboring People 1750-1800.* Ithaca and London: Cornell University Press, 1990.

Snyder, Martin P. *City of Independence: Views of Philadelphia Before 1800.* New York: Praeger Publishers, 1975.

Sparks, Jared. *The life of Benjamin Franklin: containing the autobiography, with notes and a continuation.* Boston: Tappan and Whittemore, 1853.

Spero, Patrick. *The Other Presidency: Thomas Jefferson & The American Philosophical Society.* Philadelphia: American Philosophical Society, 2018.

Talbott, Page (ed.). *Benjamin Franklin: In Search of a Better World.* New Haven, CT: Yale University Press, 2005.

Tatum, George B. *Philadelphia Georgian: The City House of Samuel Powell and Some of Its Eighteenth-Century Neighbors.* Middletown, CT: Wesleyan University Press, 1976.

Teitelman, S. Robert. *Birch's Views of Philadelphia: A Reduced Facsimile of the City of Philadelphia…. as it appeared in the Year 1800 With Photographs of the Sites in 1960 & 1982.* Philadelphia: The Free Library of Philadelphia, University of Pennsylvania Press, 1983.

Thomas, Gabriel. *A Historical and Geographical Account of the Province and Country of Pennsylvania and of West-New-Jersey in America.* London: A. Baldwin, 1698. (Reprinted from the original with Introduction by Cyrus Townsend Brady, LL.D, Cleveland: The Burrows Brothers Company, 1903.)

Thomas, Joe. *A Synopsis of the History of Moreland Township and Willow Grove.* Upper Moreland Historical Association (online PDF).

Tolles, Frederick B. *Meeting House and Counting House: The Quaker Merchants of Colonial Philadelphia 16872-1763.* New York: W.W. Norton & Company, 1948.

Treadway, Vera Thompson. *The Duffield Story: Over three hundred years of Duffield ancestry 1682–1991.* Charleston, WV: privately printed, 1991. On deposit at the Historical Society of Pennsylvania.

Van Horne, John C. *Religious Philanthropy and Colonial Slavery: The American Correspondence of the Associates of Dr. Bray, 1717–1777.* Urbana and Chicago, IL: University of Illinois Press, 1985.

_____ "The Education of African Americans in Benjamin Franklin's Philadelphia." The Good Education of Youth": *Worlds of Learning in the Age of Franklin*, ed. John H. Pollack. Oak Knoll Press, 2009, pp.73–99.

Wainwright, Nicholas B. *A Philadelphia Story: The Philadelphia Contributionship for the Insurance of Houses from Loss by Fire*. Philadelphia: Wm F. Fell Co., Printers, 1952.

_____ *Colonial Grandeur in Philadelphia: The House and Furniture of General John Cadwalader*. Philadelphia: The Historical Society of Pennsylvania, 1964.

Warner, Sam Bass Jr. *The Private City: Philadelphia in Three Periods of Its Growth*. Second Edition. Philadelphia: University of Pennsylvania Press, 1987.

Wharton, Anne Hollingsworth. *Colonial Days & Dames*. Philadelphia: J.B. Lippincott Co., 1894.

_____ *Through Colonial Doorways*. Philadelphia: J.B. Lippincott & Company, 1893.

_____ *Salons Colonial and Republican*. Philadelphia: J.B. Lippincott Company, 1900.

White, Francis. *The Philadelphia Directory*. Philadelphia: Printed by Young, Steward, and McCullough, 1785.

## PERIODICALS

Bache, Richard Meade. "The So-Called 'Franklin Prayer Book.'" Reprinted from the *Pennsylvania Magazine of History and Biography* for July, 1897.

Gillingham, Harrold E. "Early American Indian Medals," *The Magazine Antiques*, December 1924, pp.312–315.

Pennington, Reverend Edgar Legare. "The Work of the Bray Associates in Pennsylvania." *The Pennsylvania Magazine of History and Biography* (1934) Vol. LVII, No.1, pp.1–11.

Powell, Richard E. Jr. "Coachmaking in Philadelphia: George and William Hunter's Factory of the Early Federal Period." *Winterthur Portfolio*, Vol. 28, No. 4 (Winter, 1993), pp. 247–277.

Quinlan, Maurice J. "Dr. Franklin Meets Dr. Johnson", *The Pennsylvania Magazine of History and Biography*, January 1949, Vol.73, No. 1, pp.34–44.

Roach, Hannah Benner. *Colonial Philadelphians*. Philadelphia: The Genealogical Society of Pennsylvania Monograph Series No.3, 1999.

_____ *The Pennsylvania Militia in 1777*. Reprint of an article in the *PA Genealogical Magazine*, PGM Vol.23 No.3, pp.161–230.

_____ "Benjamin Franklin Slept Here," Philadelphia: *The Pennsylvania Magazine of History and Biography*, Vol.84, No.2 (April 1960), pp.127–174.

Smith, Billy G. "Death and Life in a Colonial Immigrant City: A Demographic Analysis of Philadelphia." *The Journal of Economic History"* Vol.37, No.4 (December 1977), pp.863–889.

Wainwright, Nicholas B. "Plan of Philadelphia." *The Pennsylvania Magazine of History and Biography*, Vol. 80, No. 2 (April, 1956), pp.164–226.

# COLONIAL PHILADELPHIA FURNITURE & DECORATIVE ARTS

**BOOKS**

Bjerkoe, Ethel Hall. *The Cabinetmakers of America*. Garden City, New York: Doubleday & Company, Inc., 1957.

Boudreau, George W. and Lovell, Margaretta Markle. *A Material World: Culture, Society, and The Life of Things in Early Anglo-America*. University Park, PA: The Pennsylvania State University Press, 2019.

Bridenbaugh, Carl. *The Colonial Craftsman*. New York: New York University Press, 1950.

Brinton, Mary Williams. *Their Lives and Mine*. Philadelphia: privately printed, 1972.

Brix, Maurice. *List of Philadelphia Silversmiths and Allied Artificers From 1682-1850*, Philadelphia: privately printed, 1920.

Christie's. *Fine American Furniture, Silver, Folk Art and Decorative Arts*. New York. June 2, 1990. (Lot 180)

Cooper, Wendy A. *In Praise of America: American Decorative Arts, 1650–1830: Fifty Years of Discovery Since the 1929 Girl Scouts Loan Exhibition*. New York: Alfred A. Knopf, 1980.

_____ and Minardi, Lisa. *Paint, Pattern & People: Furniture of Southeastern Pennsylvania 1725–1850*. A Winterthur Book, Distributed by The University of Pennsylvania Press, 2011, 2015.

DAPC (Decorative Arts Photographic Collection). Winterthur Museum, Garden and Library. Winterthur, DE.

Davidson, Marshall B., and Stillinger, Elizabeth. *The American Wing at the Metropolitan Museum of Art*. New York: Harrison House, 1985.

Downs, Joseph. *American Furniture: Queen Anne and Chippendale Periods in the Henry Francis du Pont Winterthur Museum*. New York: Bonanza Books, 1952.

Elder, William Voss, and Stokes, Jayne E. *American Furniture 1680-1880 From the Collection of the Baltimore Museum of Art*. Baltimore, MD: Baltimore Museum of Art, 1987.

Fales, Martha Gandy. *Joseph Richardson and Family: Philadelphia Silversmiths*. Middletown, CT: Wesleyan University Press, 1974.

Gerstell, Vivian S. *Silversmiths of Lancaster, Pennsylvania 1730-1850*. Lancaster County Historical Society, 1972.

Halsey, R.T.H. and Tower, Elizabeth. *The Homes of Our Ancestors: New England's Furniture & Interiors Through Colonial & Early Republican Times*, New York: Doubleday, Doran and Company, Inc., 1936.

Heckscher, Morrison H., and Bowman, Leslie Greene. *American Rococo, 1750-1775: Elegance in Ornament*. New York: The Metropolitan Museum of Art, 1992.

_____ *American Furniture in the Metropolitan Museum of Art II, Late Colonial Period: The Queen Anne and Chippendale Styles*. New York: The Metropolitan Museum of Art, 1985.

Holloway, Edward Stratton. *American Furniture and Decoration Colonial and Federal.* Philadelphia: J.B. Lippincott Company, 1928.

Hornor, William MacPherson Jr. *Blue Book Philadelphia Furniture William Penn to George Washington.* Philadelphia: private printing, 1935.

Hummel, Charles F. *A Winterthur Guide to Chippendale Furniture: Middle Atlantic and Southern Colonies.* New York: Crown Publishers, Inc., A Winterthur Book, 1976.

Jobe, Brock et al. *American Furniture with Related Decorative Arts 1660-1830: The Milwaukee Art Museum and the Layton Art Collection.* New York: Hudson Hills Press, 1991.

Kirtley, Alexandra Alevizatos. *The 1772 Philadelphia Furniture Price Book: An Introduction and Guide,* with facsimile of original. Philadelphia: Philadelphia Museum of Art in association with Antique Collectors' Club Ltd., 2005.

_____ *American Furniture 1650-1840: Highlights from the Philadelphia Museum of Art.* New Haven: Yale University Press, 2020.

Krill, Rosemary Troy. *Early American Decorative Arts, 1620-1860: A Handbook for Interpreters.* Roman & Littlefield (revised and enhanced edition), 2010.

Lindsey, Jack L. *Worldly Goods: The Arts of Early Pennsylvania, 1680-1758.* Philadelphia: Philadelphia Museum of Art, 1999.

Miller, Edgar G., Jr. *American Antique Furniture: A Book for Amateurs.* Volume 2. New York: Dover Publications, 1966 (republication of 1937 first edition).

Philadelphia Museum of Art. *Philadelphia: Three Centuries of American Art: Bicentennial Exhibition April 11 - October 10, 1976.* Philadelphia: 1976.

Prime, Alfred Coxe. *The Arts & Crafts in Philadelphia, Maryland, and South Carolina.* New York: Da Capo Press, 1969.

Puig, Francis J. and Conforti, Michael (ed.). *The American Craftsman and the European Tradition 1620–1820.* Hanover, NH: Distributed by University Press of New England for the Minneapolis Institute of Arts, 1989.

Sack, Israel. *American Antiques from Israel Sack Collection, Vol. 3.* Brochure 18, 1970 to Brochure 21, 1972. Washington, D.C.: Highland House Publishers Inc., 1972.

Seaman, Mary Thomas. *Thomas Richardson of South Shields, Durham County, England and His Descendants in The United States of America.* New York: private printing, 1929.

Snyder, John J. Jr. *Philadelphia Furniture & Its Makers.* Antiques Magazine Library. New York: Main Street/Universe Books, 1975.

Stiefel, Jay Robert. *The Cabinetmaker's Account: John Head's Record of Craft & Commerce in Colonial Philadelphia, 1718–1753.* Philadelphia: The American Philosophical Society Press, 2019.

Van Horn, Jennifer. *The Power of Objects in Eighteenth-century British America.* Chapel Hill: University of North Carolina Press, 2017.

Ward, Gerald W.R. (ed.). *Perspectives on American Furniture: A Winterthur Book.* New York: W.W. Norton & Company, 1988.

Zimmerman, Philip D. "Dating William Savery's Furniture Labels and Implications for Furniture History," *American Furniture 2018*. Milwaukee: Chipstone Foundation, pp.193–214.

**PERIODICALS**

Beckerdite, Luke. "Philadelphia Carving Shops" Part I, II, III. *The Magazine Antiques*, May 1984, pp.1120–1133, September 1985, pp.498–513, May 1987, pp.1044–1063.

Brazer, Clarence Wilson. "Jonathan Gostelowe: Philadelphia Cabinet and Chair Maker." *The Magazine Antiques*, June, 1926, pp. 385–395.

Hallowell, Marguerite. "Some Quaker Furniture Makers in Colonial Philadelphia," *Bulletin of Friends Historical Association*, Autumn 1958, pp.67–72.

# HOROLOGY IN EDWARD DUFFIELD'S TIME

**BOOKS**

Andrewes, William J.H. *The Quest for Longitude*. Cambridge, MA: Collection of Historical Scientific Instruments, Harvard University, 1996.

Bailey, Chris. *Two Hundred Years of American Clocks & Watches*. Englewood, NJ: A Rutledge Book, Prentice-Hall, Inc., 1975.

Baillie, G.H. *Clocks and Watches: An Historical Bibliography Volume 1*. London: The Holland Press, 1978 (reprinted from 1951 edition).

Battison, Edwin A. and Kane, Patricia E. *The American Clock 1725-1865: The Mabel Brady Garvan and Other Collections at Yale University*. Greenwich, CT: New York Graphic Society Limited, 1973.

Bedini, Silvio A. *Thinkers and Tinkers: Early American Men of Science*. New York: Charles Scribner's Sons, 1975.

_____ *Early American Scientific Instruments and Their Makers*. Washington, D.C.: Museum of History and Technology, Smithsonian Institution, 1964.

_____ *The Life of Benjamin Banneker*. New York: Charles Scribner's Sons, 1972.

Boorstin, Daniel J. *The Discoverers: A History of Man's Search to Know His World and Himself*. New York: Vintage Books, 1985.

Campbell, R. *The London Tradesman: Being a Compendious View of All the Trades, Professions, Arts, both Liberal and Mechanic, now practiced in the Cities of London and Westminster, Calculated For the Information of Parents, and Instruction of Youth in their Choice of Business*. London: printed by T. Gardner, 1747.

Clark, Victor S. *History of Manufactures in the United States 1607–1860*. Washington, D.C.: Published by the Carnegie Institution of Washington, 1916.

Cox, James. *A descriptive inventory of the several exquisite and magnificent pieces of mechanism and jewellery: comprised in the schedule annexed to an Act of Parliament, made in the thirteenth year of the reign of His Majesty, George the Third; for enabling Mr. James Cox of the city of London, Jeweler, to dispose of his museum by way of lottery.* London: 1774.

Cramer, John Andrew, M.D. *Elements Of the Art of Assaying Metals....* London: Printed for Tho. Woodward, 1741.

Craven, Maxwell. *John Whitehurst: Innovator, Scientist, Geologist and Clockmaker.* Stroud, UK: Fonthill Media Limited, 2015.

Cumming, Alexander. *The Elements of Clock and Watch-work, Adapted to Practice, In Two Essays.* London: Printed for the Author, 1766

Defoe, Daniel. *The Complete English Tradesman.* 1726.

Derham, William. *The Artificial Clock-maker, A Treatise of Watch, and Clock-work:....* London: printed for James Knapton, 1696.

_____ F.R.S. *Physico-Theology; or, A Demonstration of the Being and Attributes of GOD from his Works of Creation; with large Notes and many curious Observations.* 1732.

Distin, William H. and Bishop, Robert. *The American Clock: A Comprehensive Pictorial Survey 1723-1900.* New York: E.P. Dutton & Company, Inc., 1976.

Drepperd, Carl W. *American Clocks and Clock Makers.* Boston: Charles T. Branford Company, 1958.

Drost, William E. *Clocks and Watches of New Jersey.* Elizabeth, NJ: Engineering Publishers, 1966.

Dworetsky, Lester, and Dickstein, Robert. *Horology Americana.* Roslyn Heights, NY: Horology Americana, Inc. 1972.

Dzik, Sunny. *Engraving on English Table Clocks: Art on a Canvas of Brass 1660-1800.* Oxford, UK: Wild Boar Publications, 2019.

Eckhardt, George H. *Pennsylvania Clocks and Clockmakers: An Epic of Early American Science, Industry, and Craftsmanship.* New York: Bonanza Books, 1955.

Ferguson, James, F.R.S. *Tables and Tracts Relative to Several Arts and Sciences.* The Second Edition, With Additions. London: Printed for W. Strahan, 1771.

_____ *Astronomy explained upon sir Isaac Newton's principles....* Second edition, London, 1757. Third edition, London, 1764.

_____ *Ferguson's dissertation upon the phaenomena of the Harvest-moon; also, the description and use of a new four-wheeled orrery; and an essay upon the moon's turning round her axis, with plates.* London, 1747.

_____ *Select Mechanical Exercises: Showing how to construct different Clocks, Orreries, and Sundials....* London: Printed for W. Strahan: and T. Cadell, 1773.

Fennimore, Donald L., and Hohmann, Frank L. III. *Stretch: America's First Family of Clockmakers.* Wilmington, DE: A Winterthur Book, Hohmann Holdings LLC, 2013.

_____ *David Rittenhouse: Philosopher-Mechanick of Colonial Philadelphia and His Famous Clocks.* Wilmington, DE: Winterthur Museum, Garden, and Library, 2023.

Forman, Bruce Ross. *Clockmakers of Montgomery County 1740-1850.* Norristown, PA, 2000.

Fox, Elizabeth. *Like Clockwork: The Mechanical Ingenuity and Craftsmanship of Isaiah Lukens (1779-1846).* George Washington University: Thesis for the degree of Master of Arts in Decorative Arts & Design History, 2018.

Gaboury. *Manuel utile et curieux sur la mesure du temps: contenant des methodes tres-faciles pour pouvoir par soi-meme:…..* Angers: chez Parisot. Et se vend a Paris, chez Guillyn, 1770.

Gibbs, James W. *Pennsylvania Clocks and Watches.* University Park and London: The Pennsylvania State University Press, 1984.

_____ *Dixie Clockmakers.* Gretna, LA: Pelican Publishing Company, 1979.

Goodison, Nicholas. *Ormolu: The Work of Matthew Boulton.* London: Phaidon Press Limited, 1974.

Great Britain. *An Act for the more effectual preventing of frauds and abuses committed by persons employed in the manufacture of clocks and watches.* London: printed by Thomas Baskett, 1753.

Harris, J. Carter. *The Clock and Watch Makers American Advertiser.* Ticehurst, Wadhurst, Sussex, UK: Antiquarian Horological Society, 2003.

Harrison, John. *Remarks on a pamphlet lately published by the Rev. Mr. Maskelyne: under the authority of the Board of Longitude.* London: printed for W. Sandy, 1767.

_____ *A Description Concerning Such Mechanism As Will Afford a Nice, or True, Mensuration of Time;….* London: Printed for the Author, 1775.

_____ *The Principles of Mr. Harrison's Timekeeper, with Plates of the Same…..* London: Printed by W. Richardson and S. Clark, 1767.

Hatton, Thomas, Watch-Maker. *An Introduction to the Mechanical Part of Clock and Watch Work…* London: Printed for T. Longman, 1773.

Hawksbee, F., F.R.S. *Physico-Mechanical Experiments, on various Subjects; with Explanations of all the Machines (the Figures of which are curiously engraved on Copper) used in making the Experiments.* 1719.

Hindle, Brooke. *The Pursuit of Science in Revolutionary America 1735–1789.* Chapel Hill, NC: The University of North Carolina Press, 1956.

_____ *David Rittenhouse.* Princeton, NJ: Princeton University Press, 1964.

Hohmann, Frank L. III. *Timeless: Masterpiece American Brass Dial Clocks.* New York: Hohmann Holdings LLC, 2009.

Holder, William, D.D. *A Discourse concerning Time, with Application of the Natural Day, and Lunar Month, and Solar Year, as natural, for the better understanding of the Julian Year and Calendar, etc.* 1694. [Given by Mr. Bush.]

Hoopes, Penrose R. *Connecticut Clockmakers of the Eighteenth Century.* Rutland, VT: Charles E. Tuttle Company, 1975.

_____ *Shop Records of Daniel Burnap Clockmaker.* Hartford: The Connecticut Historical Society, 1958.

Howse, Derek. *Greenwich Time and the Longitude.* London: National Maritime Museum, 1997.

Hummel, Charles F. *With Hammer in Hand: The Dominy Craftsmen of East Hampton, New York.* Charlottesville, VA: The University Press of Virginia, Published for the Henry Francis du Pont Winterthur Museum, 1968.

Hutchins, Levi. *The Autobiography of Levi Hutchins with A Preface, Notes, and Addenda, by his Youngest Son.* Cambridge, MA: Private Edition Printed at the Riverside Press, 1865.

Huygens, Christiaan. *The Pendulum Clock or Geometrical Demonstration Concerning the Motion of Pendula as Applied to Clocks.* Paris: Apud F. Juguet, 1673.

James, Arthur E. *Chester County Clocks and Their Makers.* Eaton, PA: Schiffer Publishing, Ltd., 1947 and 1976.

LaFond, Edward Francis, Jr.. *The Henry Francis du Pont Winterthur Museum Collection of American Clocks.* Thesis, M.A. University of Delaware Winterthur Program in Early American Culture. June 1965.

_____ Unpublished draft manuscript, The Winterthur Library: Winterthur Archives, 1990.

Leadbetter, Charles. *Mechanick Dialling Or, the New Art of Shadows:…* London: printed for W. Wicesteed, 1756.

Leybourn, William. *Dialing: Shewing how to make all Sorts of Dials, and to adorn them with all useful Furniture relating to the Course of the Sun.* London, 1682.

Loomes, Brian. *Grandfather Clocks and Their Cases.* New York: Arco Publishing, Inc., 1985.

Martin, Benjamin. *The Young Student's Memorial Book, or Pocket Library: containing… Dialling… Mechanics ….* London: 1736.

Mayr, Otto. *Authority, Liberty & Automatic Machinery in Early Modern Europe.* Baltimore: The Johns Hopkins University Press, 1986.

McEvoy, Rory and Betts, Jonathan (eds.). *Harrison Decoded: Towards a Perfect Pendulum Clock.* Greenwich, England: Royal Observatory Greenwich and Oxford University Press, 2020.

McGraw, Judith A. (ed.). *Early American Technology: Making & Doing Things From the Colonial Era to 1850.* Chapel Hill, NC: University of North Carolina Press, 1994.

Montgomery, Charles F. *The American Clock 1725–1865* (Two Yale Student Exhibitions, May 18–June 28, 1974). New Haven: Garvan Furniture Study, 1974.

Moore, N. Hudson. *The Old Clock Book.* New York: Tudor Publishing Company, 1911.

Moraley, William. *The Infortunate: The Voyage and Adventures of William Moraley, an Indentured Servant.* Second Edition. University Park, PA: The Pennsylvania State University Press, 2005.

Morris, Philip E. Jr. *American Wooden Movement Tall Clocks 1712-1835.* Hoover, AL: Heritage Park Publishing, 2011.

Multhauf, Robert P. *Catalogue of Instruments and Models in the Possession of the American Philosophical Society.* Philadelphia: The American Philosophical Society, 1961.

Mudge, Thomas. *Thoughts on the means of improving watches: and more particularly those for the use of the sea. Partly deduced from reason, and partly from the observation of effects attributed to particular causes.* London: 1765.

Olton, Charles S. *Artisans for Independence: Philadelphia Mechanics and the American Revolution.* Syracuse, NY: Syracuse University Press, 1975.

_____ *Philadelphia Artisans and the American Revolution.* University of California, Berkeley, unpublished Ph.D. thesis, 1967.

Palmer, Brooks. *The Book of American Clocks.* New York: The MacMillan Company, 1950.

Petrucelli, Steven P. *Striking Beauty: New Jersey Tall Case Clocks: 1730–1830.* Cranbury, NJ: Adams Brown Company, Inc., 2023.

Philadelphia Antiques Show. *It's About Time: Loan Exhibit - 2000.* Robert and Katharine Booth, Curators. Philadelphia, 2000.

Prager, Frank D. (ed.). *The Autobiography of John Fitch.* Philadelphia: The American Philosophical Society, 1976.

Pryce, W.T.R and Davies, T. Alun. *Samuel Roberts, Clock Maker: An eighteenth-century craftsman in a Welch rural community.* National Museum of Wales, Welsh Folk Museum, 1985.

Quimby, Ian M.G. "Edward Duffield: Artisan Gentleman." Unpublished manuscript, paper for Art-506, University of Delaware, Winterthur: January 10, 1963.

_____ (ed.). *The Craftsman in Early America.* A Winterthur Book. New York: W.W. Norton & Company, 1984.

_____ *Apprenticeship in Colonial Philadelphia.* New York: Garland Publishing, 1985.

Rittenhouse, David and Ewing, John. *1769 Transit of Venus.* American Philosophical Society Essays. Philadelphia: American Philosophical Society, 2012.

Roberts, Derick. *The Bracket Clock.* London: David & Charles, 1982.

Robey, John. *The Longcase Clock Reference Book, Second Edition, Two Volumes.* Mayfield, Ashbourne, Derbyshire, UK: Mayfield Books, 2013.

_____ *Gothic Clocks to Lantern Clocks: Short-Duration Clocks & Rural Clocks 1480-1800.* Mayfield, Ashbourne, Derbyshire, UK: Mayfield Books, 2021.

Robinson, Tom. *The Longcase Clock.* UK: Antique Collectors' Club, 1995 Revised Edition.

Royal Society. *Philosophical Transactions, Giving Some Account of the Ingenious in Many Considerable Parts of the World.* Volume XLVII. For the Years 1751 and 1752. London: Printed for C. Davis, Printer to the Royal Society, M.DCC.LIII.

Shaffer, Douglas H. *Clocks.* The Smithsonian Illustrated Library of Antiques. New York: Cooper-Hewitt Museum, 1980.

Shelley, Frederick. *Early American Tower Clocks.* Columbia, PA: National Association of Watch and Clock Collectors, 1999.

Smart, Charles E. *The Makers of Surveying Instruments in America Since 1700*, Troy, NY: Regal Art Press, 1962.

Smith, Alan. *A Catalogue of Tools for Watch and Clock Makers by John Wyke of Liverpool.* Charlottesville, VA: University Press of Virginia for the Henry Francis du Pont Winterthur Museum, 1978.

Smith, John. *"A table of equations, for reducing the unequality of natural days to a mean and equal time: Designed chiefly in order to the more true adjusting, and right managing of pendulum clocks and watches.* London: 1 sheet, 1686. Ben Franklin's copy bound in a volume, number 54 in the contents list.

J.S. Clock-maker. *Horological Dialogues In Three Parts Shewing The Nature, Use, and right Managing of Clocks and Watches:…..* London: Printed for Jonathan Edwin, 1675.

Smith, Murphy D. *Due Reverence: Antiques in the Possession of the American Philosophical Society.* Philadelphia: The American Philosophical Society, 1992.

Sotheby Parke Bernet Inc. *The American Heritage Society Auction of Americana.* New York. November 16-18, 1972, Lot 676.

Spittler, Sonya L. and Thomas J., and Bailey, Chris H. *Clockmakers and Watchmakers of America by Name and by Place, Second Edition.* Columbia, PA: National Association of Watch & Clock Collectors, 2011.

Stearns, Raymond Phineas. *Science in the British Colonies of America.* Urbana: University of Illinois Press, 1970.

Sullivan, Gary R. and Van Winkle Keller, Kate. *Musical Clocks of Early America 1730–1830.* North Grafton, MA: The Willard House & Clock Museum, 2017.

Tennant, M.F. *Longcase Painted Dials: Their History and Restoration.* London: N.A.G. Press, 1995.

Thomson, Richard. *Antique American Clocks & Watches.* New York: Van Nostrand Reinhold Company, 1968.

Town, Edward & McShane, Angela (ed.). *Marking Time: Objects, People, and Their Lives, 1500–1800.* New Haven: Yale Center for British Art, Yale University Press, 2020.

Whisker, James Biser. *Pennsylvania Clockmakers, Watchmakers, and Allied Crafts.* Cranbury, NJ: Adams Brown Co., 1990.

_____ *Pennsylvania Clockmakers and Watchmakers, 1660–1900.* Lewiston, NY: The Edwin Mellen Press, 1996.

Winchester, Alice (ed.). *The Antiques Treasury of Furniture and Other Decorative Arts.* New York: Galahad Books, 1959.

*The Complete Tradesman. 2 Vols. Treating of the several Points necessary to be known, both by the younger and more experienced Tradesman; with regard to Diligence, Over-Trading, expensive Living, too early Marrying, Diversion, Credit, Partnership, Punctuality, Projects, Engrossing, etc.* 1732.

## PERIODICALS

Barr, Lockwood. "English Clocks in American Cases," *American Collector*, Volume IX, Number 11 (December 1940).

Bassett, Lynne Zacek. "Woven Bead Chains of the 1830s," *The Magazine Antiques*, December 1995, pp.799–807.

Cheney, Robert C. "Roxbury Eight-Day Movements and the English Connection, 1785–1825," *The Magazine Antiques*, April 2000, pp.606–613.

Eckhardt, George H. "Edward Duffield, Benjamin Franklin's Clockmaker," *The Magazine Antiques*, March 1960, pp.284–286.

_____ "Franklin's Clock Comes Home," *The Philadelphia Inquirer*, May 31, 1959, p.132.

Fales, Martha Gandy. "Thomas Wagstaffe, London Quaker Clockmaker," *The Connoisseur* (November 1962): pp.193–201.

Frishman, Bob. "Horology in Art, Part 17" (Portrait of Johannes Kelpius by Dr. Christopher Witt). *NAWCC Watch & Clock Bulletin*, November/December 2014, p.583.

Kernodle, George H. "Concerning the Simon Willard Legacy," *The Magazine Antiques*, June 1952, pp.523–8.

LaFond, Edward Francis, Jr. "Isaac Heron: The Outspoken Clockmaker," Columbia, PA: *NAWCC Watch and Clock Bulletin*, June 1979, pp.291–307.

Newman, Richard. "Colonial and Early American Watchmakers." Columbia, PA: *NAWCC Watch and Clock Bulletin*, December 2010, pp.692–706.

Olton, Charles S. "Philadelphia's Mechanics in the First Decade of Revolution 1765–1775." *The Journal of American History*. Oxford University Press, Vol. 59, No. 2 (September 1972), pp.311–326.

Sampson, Paul E. "The Cosmos in a Cabinet: Performance, Politics, and Mechanical Philosophy in Henry Bridges' 'Microcosm.'" *Endeavor* 43, 2019, pp.25–31.

Sperling, David. "A Closer Look at Edward Duffield: Philadelphia Clockmaker." Columbia, PA: *NAWCC Bulletin*, October 2006, pp.589–593.

_____ "Great Treasure In A Small Museum: The Life and Work of Clockmaker Edward Duffield." *Antiques and The Arts Weekly*, January 6, 2006, pp.76–77.

Stiefel, Jay Robert. "'A Clock for the Rooms': The Horological Legacy of the Library Company of Philadelphia." *Antiquarian Horology* (December 2006): pp.804–826.

_____ "The Library Company of Philadelphia: The Clocks in the Collection." *The Magazine Antiques*, August 2006, pp.88–93.

Stretch, Carolyn Wood. "Early Colonial Clockmakers in Philadelphia." *Pennsylvania Magazine of History and Biography* (June 30, 1932): pp. 225–235.

Winchester, Alice. Introductory essay, *The Magazine Antiques*, June 1952, p.507.

Wood, Stacy B.C., Jr. "Rudy Stoner 1728–1769 Early Lancaster, Pennsylvania, Clockmaker," *Bulletin* of the National Association of Watch and Clock Collectors, Inc., Columbia, PA, February 1977, pp.21–32.

# 𝔥𝔬𝔯𝔬𝔩𝔬𝔤𝔦𝔠𝔞𝔩 𝔇𝔦𝔰𝔮𝔲𝔦𝔰𝔦𝔱𝔦𝔬𝔫𝔰

Concerning the

# NATURE of TIME,

AND THE

## 𝔑𝔢𝔞𝔰𝔬𝔫𝔰 why all 𝔇𝔞𝔶𝔰, from 𝔑𝔬𝔬𝔫 to 𝔑𝔬𝔬𝔫, are not alike Twenty Four Hours long.

In which appears the Impossibility of a Clock's being always kept exactly true to the Sun.

With TABLES of EQUATION and Newer and Better RULES than any yet extant, how thereby precisely to adjust ROYAL PENDULUMS, and keep them afterwards, as near as possible to the apparent Time.

With a TABLE of PENDULUMS, shewing the BEATS that any Length makes in an Hour.

A Work very necessay for all that would understand the true way of rightly managing Clocks and Watches.

By *JOHN SMITH*, C.M.

To which is added,
The best Rules for the Ordering and Use both of the 𝔔𝔲𝔦𝔠𝔨-𝔖𝔦𝔩𝔳𝔢𝔯 and 𝔖𝔭𝔦𝔯𝔦𝔱 𝔚𝔢𝔞𝔱𝔥𝔢𝔯-𝔊𝔩𝔞𝔰𝔰𝔢𝔰: And Mr. *S. Watson's* Rules for adjusting a Clock by the *Fixed Stars*.

LONDON: Printed for *Richard Cumberland* at the *Angel* in S. Paul's Church-Yard. 1694.

1694 technical horology book Duffield may have owned and studied. Author collection and photo.

# ENDNOTES

1. Catalogue No.3, APS #58.62.
2. Fairbanks, Jonathan, letter to Mrs. Donald L. Symington, March 21, 1975.
3. Hummel, p.129. Italics are author's.
4. Eckhardt, p.31.
5. Founded in 1688, this oldest-surviving Baptist church in Pennsylvania now is a national and Baptist landmark.
6. President of the Free Society of Traders, Dr. More later became the first Chief Justice of Pennsylvania.
7. Neill 1, 1875, p.67.
8. *Lawmaking and Legislators in Pennsylvania, A Biographical Dictionary*, Volume Two, p.334.
9. Pennsylvania DAR GRC Report, GRC-PA:s1:v10, biblerecords bk. 1/Philadelphia Chapter - Page 54.
10. Cohen, p.60.
11. Labaree Vol. 9, p.293.
12. University of Pennsylvania, University Archives.
13. Neill 1, p.67.
14. Accession No. 33.120.350a,b.
15. Neill 1, p.70.
16. *Pennsylvania Journal*, July 7, 1748 p.4.
17. This miniature sold in 2014 at the Delaware Antiques Show. No similar miniatures of Edward or other family members have appeared.
18. *Pennsylvania Gazette*, July 25, 1765, p.1.
19. Labaree Vol. 13, p.198.
20. *Pennsylvania Gazette*, April 17, 1766, p.1.
21. Neill 1, p.79.
22. Neill 1, p.76.
23. Labaree Vol. 21, pp.268–269.
24. *Pennsylvania Gazette*, August 30, 1770, p.5.
25. Neill 1, pp.79–80.
26. Graham 2, p.43.
27. Franklin had previously provided the young doctor with a letter of introduction to Lord Kames in Edinburgh.
28. Neill 1, p.91.
29. http://tinyurl.com/32r4ryzt.
30. Copy of original indenture provided to author by Menkevich.
31. *Claypoole's American Daily Advertiser*, October, 13, 1796, p.3.
32. *Claypoole's American Daily Advertiser*, February 10, 1797, p.2. Also Drinker p.742.
33. Pottsgrove Manor Archives.
34. Letter dated October 9, 1897, in *Life and Correspondence of James Iredell, One of the Associate Justices of the Supreme Court of the United States*, Griffith J. McRee, 1858, letter dated October 9, 1897.Vol.II, p.491.
35. *Life and Correspondence of James Iredell, One of the Associate Justices of the Supreme Court of the United States*, Griffith J. McRee, 1858, letter dated October 10, 1793, Vol.II,p.402.
36. *True American and Commercial Daily Advertiser*, January 22, 1800, p.1.
37. Boston Athenaeum 926.P75.n.

38. Boston Athenaeum 929.N31.
39. Philadelphia Bulletin, November 6, 1854.
40. British Library, 1447.i.7.
41. *Pennsylvania Gazette*, June 24, 1756, p.3.
42. *Pennsylvania Gazette,* November 8, 1770 page 4.
43. Historical Society of Pennsylvania, Frank M. Etting Collection (0193), Box 47, Misc. Manuscripts V.1, p.123.
44. PMHB X (1886) pp.461-2, Quimby p.13.
45. "John Cairns (1751–1809) and Other Early American Watchmakers," David Cooper, *NAWCC Bulletin*, February 2002, pp.26–38.
46. *The Providence Gazette*, January 4, 1800.
47. Eckhardt, p.75.
48. Temple University Library P154025B.
49. Smith, Murphy, pp.2–3.
50. Rittenhouse & Ewing, p.82.
51. Smith, Murphy, p.3.
52. *Pennsylvania Gazette*, June 14, 1770, p.5.
53. Documents relating to the colonial Revolutionary and post-Revolutionary history of the State of New Jersey Vol.27 published 1880, p.437.
54. Tolles, p.260.
55. Hazard, p.194.
56. Hazard, p.574.
57. A few facts and traditions about the Lower Dublin Township, 1911.
58. Neill 1, p.72.
59. Quimby, p.12.
60. Quimby, p.12.
61. *Pennsylvania Gazette*, October 14, 1762, p.4.
62. 1774 Votes of Assembly p.7177.
63. General Braddock's Orderly Book, Library of Congress, p.XV reports: "One corporal and eight men of the Line to attend at 6 O'clock every morning, to assist the Engineers in Surveying." One of these men may have been Edward, who may have joined them at the start of the expedition, when a small portion of Braddock's army marched to Pennypacker's Mills near Philadelphia to join contingents coming from Virginia and Maryland. If he did subsequently witness the debacle and barely avoided being killed, as did twenty-three-year-old Lieutenant-Colonel George Washington, this may account for his later paying a fine to avoid military service in the Revolution. However, Edward's presence there is not confirmed in any existing records and there were no Pennsylvania troops involved, only teamsters recruited by Franklin at his own expense. Franklin's brother-in-law, John Read, was one of those wagon masters, however. He was acquainted with his one-time neighbor, Edward, and may have brought him along.
64. Martindale, p.305.
65. Labaree Vol.7 pp.208-213.
66. *Federal Gazette*, April 29, 1790, p.3.
67. Accession No.23/9269.
68. A 1929 privately-printed genealogy by a descendant of silversmith Joseph Richardson captioned a photo of a desk allegedly made by Edward Duffield68 but this reflects family lore from long ago. More plausible is that Edward sold or traded the desk to Richardson.
69. Bridenbaugh, p.272.
70. *Dunlap's Pennsylvania Packet or, the General Advertiser*, April 24, 1775, p.1.
71. Neill 1, p.73.
72. The original of the inventory is lost but a handwritten transcription was included by Ian M.G. Quimby in his January,1963, unpublished treatise, *Edward Duffield: Artisan Gentleman*. A typed transcription of the entire Quimby inventory is in the appendix.

73. Hoopes, p.49.
74. Hoopes, p.57.
75. Hoopes, p.87.
76. Cheney, p.609.
77. Lindsey, p.5.
78. Fennimore and Hohmann, p.96.
79. Eckhardt, pp.284-285.
80. Bedini, p.211.
81. Bedini, pp.199-200.
82. *Pennsylvania Gazette*, August 8, 1751, p.3.
83. *Pennsylvania Gazette*, July 17, 1760, p.3.
84. *Pennsylvania Gazette*, September 6, 1764, p.1.
85. *Pennsylvania Gazette*, March 25, 1762, p.1.
86. Willard House & Clock Museum, #2.96.12.
87. Wiederseim Associates, February 15, 2023, Lot 147.
88. For more on this clock see Chapter Nine.
89. Bonhams London, July 13, 2023, Lot 77.
90. Pryce, p.205.
91. Lindsey, p.136.
92. Scharf, p.150.
93. Tolles, p.41.
94. Hornor, p.22.
95. Scharf, p.193–4.
96. Fennimore and Hohmann, p.95.
97. Carter March & Co., "The John C Taylor Collection Part I," 2021, Exhibit No.38.
98. Milley, p.124.
99. Baillie, p.150.
100. Stiefel, pp.219, 222, and 205.
101. The Philadelphia Contributionship for the Insurance of Houses from Loss by Fire, Local ID Number S00024–002.
102. Fennimore and Hohmann, p.91.
103. Fennimore and Hohmann, p.102.
104. Robey 1, p.61–62.
105. Fennimore and Hohmann, p.96.
106. Fennimore and Hohmann, p.96.
107. Fales, pp.23 & 275.
108. Fennimore and Hohmann, p.95.
109. Unfortunately, two years later, Bagnall suffered serious financial difficulties and his household goods, including, "a sett of watchmaker's tools…" were auctioned at his dwelling on Front Street.
110. *Pennsylvania Gazette*, May 28, 1752, p.4.
111. Jared Sparks Collection of Franklin letters, Harvard University, August 6, 1759.
112. *Pennsylvania Gazette*, April 2, 1754, p.3.
113. *Pennsylvania Gazette*, July 25, 1754, p.3.
114. *Pennsylvania Gazette*, September 18, 1760 p.4.
115. *Pennsylvania Gazette*, November 20, 1760 p.4.

116. *Pennsylvania Gazette*, January 3, 1760, p.3.

117. Fales, p.219.

118. Winterthur Archives.

119. Winterthur Archives.

120. Winterthur Archives.

121. Hornor, p.129.

122. *Pennsylvania Gazette*, March 25, 1762.

123. *Pennsylvania Gazette*, December 1, 1764, p.4.

124. "Heads and dolphins" were standard dial spandrels affixed to corners and arches.

125. *Pennsylvania Gazette*, March 17, 1763, p.3.

126. *Pennsylvania Gazette*, July 14, 1763 p.4.

127. *Pennsylvania Gazette*, February 9, 1764, p.4.

128. *Maryland Gazette*, April 12, 1764, p.4.

129. *Maryland Gazette*, August 30, 1764, p.1.

130. *Pennsylvania Gazette*, November 15,1764, p.3.

131. *Pennsylvania Gazette*, July 4, 1765, p.1.

132. *Pennsylvania Gazette*, January 22, 1767, p.4.

133. *Pennsylvania Packet*, February 23, 1768, p.3.

134. *Pennsylvania Gazette*, May 5, 1768, p.8; also May 12, 1768, p.8.

135. Winterthur Museum Object Number 1961.0516.

136. Freeman's Books and Manuscripts, November 16, 2023, Lot 147.

137. *Pennsylvania Gazette*, May 7, 1772, p.3.

138. *Pennsylvania Chronicle*, May 9, 1768.

139. *Pennsylvania Gazette*, October 24, 1771, p.1.

140. *Pennsylvania Gazette*, June 27, 1771, p.3.

141. Pennsylvania Packet August 23, 1773, p.6.

142. Olton, p.3.

143. Bell 1, p.293n.

144. *Pennsylvania Gazette*, February 23, 1769, p.3.

145. *Pennsylvania Gazette*, January 4, 1770, p.4.

146. *Pennsylvania Gazette*, April 18, 1771, p.4.

147. *Pennsylvania Packet*, November 11, 1771, p.3.

148. *Pennsylvania Gazette*, May 5, 1773, p.4.

149. *Pennsylvania Packet*, January 18, 1773.

150. *Pennsylvania Packet*, October 24, 1774, p.6.

151. For more on Birnie see: "Laurence Birnie and the Apprentice Griffith Owen," Bruce Ross Forman, *NAWCC Watch & Clock Bulletin*, March/April 2015, pp.170–171. "Laurence Birnie, An Antrim Clockmaker," Killian Robinson and Bruce Forman, *NAWCC Watch & Clock Bulletin*, November/December 2015, pp.571–575.

152. *Pennsylvania Packet*, July 1 and July 11, 1780, p.1 and p.4.

153. *Independent Gazetteer*, May 25, 1782, p.3.

154. Historical Society of Pennsylvania Amb.5865.

155. Fales, pp.158 and 302.

156. *Pennsylvania Packet*, June 5, 1783, p.4.

157. *Pennsylvania Packet*, June 16, 1787.

158. Tennant, p.160.

159. Winterthur Archive TS1035 W82 TC.

160. Boorstin, p.69.

161. *Pennsylvania Packet*, May 18, 1786, p.4.

162. *Pennsylvania Gazette*, April 18 & 25, May 2 & 30, 1734.

163. *Pennsylvania Gazette*, July 19, 1737, p.3.

164. *Pennsylvania Gazette*, September 11, 1755, p.5.

165. *Pennsylvania Gazette*, May 11, 1769, p.7.

166. *Pennsylvania Gazette*, June 4, 1767, p.4.

167. *Pennsylvania Gazette*, September 5, 1771, p.1.

168. *Pennsylvania Ledger*, February 14, 1778, p.3.

169. *Pennsylvania Packet*, April 1, 1779, p.3.

170. *Pennsylvania Packet*, October 6, 1778.

171. *Pennsylvania Packet*, October 6, 1778.

172. *Pennsylvania Packet*, October 10, 1778, p.3.

173. Roberts, p.72.

174. Hiltzheimer, p.243: "Today the well-known David Rittenhouse was buried under a small building in the rear of his house, Northwest corner Seventh and Arch Streets." He was eulogized in December on behalf of the American Philosophical Society by Dr. Benjamin Rush at the First Presbyterian Church. George Washington attended as one of his final public appearances as President. Edward may have been present as well. Hiltzheimer later recorded that "At noon met Thomas and Norton Pryor; went with them to the late David Rittenhouse's observatory, to set the time piece there, which they have done since the death of that great man."

175. Fales, p.59.

176. Fales, p.297.

177. *Pennsylvania Gazette*, April 23, 1752, p.6.

178. *Pennsylvania Gazette* October 13, 1757 p.4.

179. *American Weekly Mercury*, November 16-23, 1738.

180. *Pennsylvania Gazette*, July 18, 1754, p.6.

181. Hornor, p.127.

182. Horner, p.286.

183. *Pennsylvania Gazette*, September 24, p.5.

184. *Pennsylvania Gazette*, May 12, p.8.

185. Lyons, p.147.

186. Scharf, p.263.

187. *Pennsylvania Packet*, May 27, 1785, p.3.

188. *Independent Gazetteer*, January 24, 1783, p.4.

189. *Pennsylvania Packet*, July 5, 1783, p.2.

190. *Independent Gazetteer*, January 1, 1785, p.4.

191. Hornor, p.125.

192. *Pennsylvania Chronicle*, March 27, 1769, p.71.

193. *Pennsylvania Packet*, March 9, 1779, p.3.

194. *Pennsylvania Gazette*, September 15, 1737, p.4.

195. *American Weekly Mercury*, July 6, 1738.

196. *Pennsylvania Gazette*, October 5, 1752, p.4.

197. *Pennsylvania Gazette*, April 12, 1770, p.3.

198. *Pennsylvania Gazette*, November 20, 1766, p.4.

199. *Pennsylvania Gazette*, September 24, 1767, p.3.

200. *Pennsylvania Gazette*, December 31, 1767, p.4.

201. "Thomas Wagstaffe Quaker Clock-maker" by George E. Moore, M.D. and Arthur E. James. In *Bulletin of the National Association of Watch and Clock Collectors, Inc.*, December, 1976, p.533.

202. Hornor, p.304.

203. Milnor, pp.401–402.

204. *Pennsylvania Gazette*, August 16, 1744, p.3.

205. *Pennsylvania Gazette*, December 18, 1755, p.3.

206. British Museum number 1958, 1006.2101.

207. Sampson, p.25.

208. Smith, Murphy, pp.23–24.

209. Rittenhouse and Ewing, p.26.

210. *Pennsylvania Gazette*, April 26, 1770, p.3.

211. Ford, Paul Leicester, The Works of Thomas Jefferson, Correspondence, pp.42–43.

212. Founders Online, National Archives and Records Administration, July 8, 1787.

213. *Aurora General Advertiser* notices, March 11–April 22, 1794.

214. Hazard, p.404.

215. *Pennsylvania Chronicle*, July 10, 1769, p.200.

216. *The Wager* by David Grann (New York: Doubleday 2023) tells the full story of this calamitous British naval disaster and subsequent Admiralty court-martial trials in London. Much of Grann's book is based on Bulkeley's skewed record of the grim events. A fictional account was also the subject of one of Patrick O'Brian's first seafaring novels, *The Unknown Shore* (London: Rupert Hart-Davis, 1959).

217. *Pennsylvania Gazette*, 1756, July 29 p.4, August 12 p.3, August 26 p.4.

218. *Pennsylvania Gazette*, March 10, 1757, p.3.

219. Heritage Auctions, November 11, 2011, Lot 92005.

220. Benjamin Franklin, *Political, Miscellaneous, and Philosophical Pieces*, ed. Benjamin Vaughan, (London, 1779), pp. 533–6.

221. Scharf, p.255.

222. Horle, p.1071.

223. Vestry minutes, Christ Church and St. Peter's Church, v. 2, 1761–1784

224. *Pennsylvania Journal, or, Weekly Advertiser*, July 5, 1764, p.1.

225. *Pennsylvania Gazette*, January 31, 1765, p.3.

226. Bridenbaugh, p.229.

227. *Pennsylvania Gazette*, April 24, 1766, p.5.

228. Christ Church Seating Plan, circa 1760, Greater Philadelphia GeoHistory Network.

229. Christ Church Seating Plan, circa 1760, Greater Philadelphia GeoHistory Network.

230. Vestry minutes, Christ Church and St. Peter's Church, v. 2, 1761–1784.

231. *Pennsylvania Gazette*, July 26, 1759 p.6.

232. Van Horne 1, pp.313–314.

233. Labaree Vol. 14, p.340.

234. Labaree Vol. 15, p.87.

235. Minute Book: Library Company of Philadelphia, 1768–1785, pp.53, 55.

236. *Pennsylvania Gazette*, January 23, 1772, p.5.

237. *Pennsylvania Chronicle*, October 4, 1753, p.3.

238. Van Horne 2, p.85.

239. Van Horne 2, p.318.
240. Van Horne 2, p.85.
241. Historical Society of Pennsylvania (original document dated October 13, 1774).
242. *Pennsylvania Ledger*, February 18, 1778, p.3.
243. *New-York Gazette, and Weekly Mercury*, 1778, p.2.
244. Bell & Griffenstein, p.553.
245. Bell 3, p.553.
246. Labaree Vol.26, p.488.
247. https://pspaonline.com.
248. https://archive.org/details/minutesofphilade00phil_0/page/120/mode/2up/search/Duffield.
249. Hiltzheimer, p.75.
250. Hiltzheimer, p.100.
251. Hiltzheimer, p.100.
252. Hiltzheimer, p.115.
253. Hiltzheimer, p.160.
254. Menkevich, p.25.
255. Private email to author.
256. Hiltzheimer, p.226.
257. Peters, pp.44–48.
258. Mittelberger, pp.88–89.
259. *The Magazine Antiques*, February 1954, p.392.
260. Historical Society of Pennsylvania, The Graisbury Ledger in the Reed and Forde Papers (0541) pp.72 and 127.
261. McBroom, Ann. "Daniel Quare (1648–1724): Yorkshireman, Activist, Quaker Horologist and Businessman." In *Antiquarian Horology*, March 2023, p.63.
262. Lindsey, p.30.
263. Tolles, p.64.
264. Roach 1, p.106.
265. *History Today*, September, 2008, p.26.
266. Hornor, p.21.
267. Hornor, p.57.
268. Thomas, p.43.
269. Eberlein, pp.193–194.
270. Eberlein, p.196.
271. *Hamilton's Itinerarium: Being a narrative of a journey from Annapolis, Maryland, through Delaware, Pennsylvania, New York New Jersey, Connecticut, Rhode Island, Massachusetts and New Hampshire, from May to September, 1744.*
272. Hamilton, p.19.
273. *Pennsylvania Gazette*, July 1, 1731, p.3.
274. Kelley, p.50.
275. Scharf, p.213.
276. *Pennsylvania Gazette*, October 13, 1748, p.2.
277. *Pennsylvania Gazette*, June 29, 1758, p.3.
278. Mittelberger, p.42.
279. Mittelberger, p.50.
280. Mittelberger, p.91.

281. Mittelberger, p.48.

282. Mittelberger, p.92.

283. *Pennsylvania Gazette*, December 18 & 25, p.3.

284. *Pennsylvania Gazette*, December 19, p.2.

285. *Pennsylvania Gazette*, February 7, 1765, p.1 and February 14, p.3.

286. Franklin 2.

287. *Pennsylvania Gazette*, May 29, 1755, p.3.

288. Bridenbaugh 2, p.164.

289. Franklin, Benjamin, *Letters relating to a Transit of Mercury over the Sun, which is to happen May 6th, 1753*.

290. Labaree Vol. 4, p.212.

291. Cope, Thomas D. *John Shelton's Clock Used by Mason and Dixon at Brandywine*, Proceedings of the Pennsylvania Academy of Science, 1945, Vol, XIX, pp.79–86.

292. Eberlein, p.25.

293. Davies, p.24.

294. *The Magazine Antiques*, January 1974, p.120.

295. *Pennsylvania Gazette*, December 19, 1759, p.2.

296. Franklin 3, pp.146–148.

297. Hindle, p.22.

298. *Pennsylvania Packet*, February 3, 1772, p.2.

299. Hindle 2, Chapter VIII.

300. *Pennsylvania Gazette*, June 15, 1774, p.3; Sharf p.290.

301. https://en.wikipedia.org/wiki/Israel_Bissell.

302. Lancaster, p.151.

303. Drost, p.16.

304. Martindale, p.305.

305. Literary and Philosophical Society of Manchester *Memoirs*, ii(1785), pp.357–361.

306. Harris, Carter. "A Philadelphia Clockmakers Company: Some Documentary Evidence", *NAWCC Bulletin*, December 1984, p.699.

307. Carey, Mathew.

308. Carey, p.79.

309. Powell, p.72.

310. Corner, George W., *The Autobiography of Benjamin Rush*, Published for the American Philosophical Society by Princeton University Press, 1948, p.355.

311. Drinker, p.155.

312. Powell, J.H., p.24.

313. Drinker, 1802 p.xvii.

314. Accession No. 33.120.350a,b.

315. Bulletin of the Metropolitan Museum of Art, June 1922, p.141.

316. Roach 1.

317. The Philadelphia Contributionship for the Insurance of Houses from Loss by Fire. Digital Archives. Insurance Survey S00410. Local ID Number: S00411-001A (handwritten document transcribed by author).

318. *Pennsylvania Gazette*, October 18, 1764, p. 4.

319. Labaree Vol. 12, p.348.

320. Records of the Land Office, Patent Index, A and AA Series, 1684–1781. {series #17.147}.

321. *Pennsylvania Packet*, June 2, 1781, p.4.

322. Insurance Survey S02130.

323. Martindale, p.305.

324. Martindale, p.196.

325. Mease Vol.2, p.484.

326. Dolin, p.144 & 155.

327. Document image courtesy of Joseph J. Menkevich.

328. Document image courtesy of Torben Jenk.

329. Document image courtesy of Torben Jenk.

330. Document image courtesy of Torben Jenk.

331. Almanacs Pa. M170 1801.

332. *Pennsylvania Gazette*, March 29, 1803, p.4.

333. A full illustrated description of this simplified clock movement is in *Life of James Ferguson, F.R.S.* by E. Henderson, Ll.D., published in London in 1867.

334. Labaree Vol.1, pp.237–240.

335. Benjamin Franklin, *A Proposal for Promoting Useful Knowledge among the British Plantations in America*, May 14, 1743.

336. Neill 1, p.73.

337. https://diglib.amphilsoc.org/franklindata/database

338. Cohen, p.91.

339. Labaree Vol.7, p.100.

340. Andrewes, p.207.

341. Labaree Vol.7, p.219.

342. Labaree Vol.7, p.278.

343. Labaree Vol.7, p.369.

344. Roach, p.168.

345. Labaree Vol. 10, pp.248–249.

346. Labaree Vol. 10, p.226.

347. Labaree Vol. 10, p.300.

348. Labaree Vol. 12, p.42.

349. Eisenhart, p.153.

350. Labaree Vol. 14, p.192

351. Labaree Vol. 14, p.280

352. Labaree Vol. 15, p.46.

353. Labaree Vol. 15, p.291.

354. Labaree Vol. 15, p.186.

355. Labaree Vol. 15, p.291.

356. Labaree Vol.18, pp.190–191.

357. Labaree Vol.18, pp.194–194.

358. Labaree Vol.19, pp.149–150.

359. Labaree Vol.19, pp.229–230.

360. Christie's June 12, 2019. (Original letter).

361. Labaree Vol. 22, p.491.

362. Labaree Vol. 22, p.566.

363. Labaree Vol. 30, p.334.

364. Labaree Vol. 33, p.423.

365. Labaree Vol. 29, p.615.
366. Labaree Vol. 30, p.352.
367. Labaree Vol. 30, p.551.
368. Labaree Vol. 28, p.392.
369. Labaree Vol. 33, p.99.
370. Labaree Vol. 33, p.271.
371. Labaree Vol. 35, p.278.
372. Temple University Library P154025B.
373. Craven, p.115.
374. Sparks, pp.519–23.
375. Sparks, pp. 521–522.
376. *Pennsylvania Gazette*, April 21, 1790, p.2.
377. *Thomas's Massachusetts Spy, or The Worcester Gazette*, June 24, 1790, p.3.
378. Talbot, p.179, Fig.5.19.
379. Neill 1, p.73.
380. Via Dr. Richard Mones' notes.
381. *Pennsylvania Packet, and Daily Advertiser*, October 21, 1790, p.1.
382. *Aurora General Advertiser*, May 24, 1792, p.2.
383. Hart, Charles Henry et al, "The Wilson Portrait of Franklin: Earl Grey's Gift to the Nation," *The Pennsylvania Magazine of History and Biography*, Vol.30, No.4, p.412.
384. *Pennsylvania Gazette*, April 22, 1762 p.4.
385. *Pennsylvania Gazette*, April 23, 1762 p.4.
386. *Pennsylvania Gazette*, December 25,1766 p.3 and January 8, 1767 p.5.
387. *Pennsylvania Gazette*, August 14, 1740, p.7.
388. American Philosophical Society, 58-32, Multhauf p.15.
389. Hindle, p.246.
390. Petrucelli, p.61.
391. Library Company of Philadelphia, Books & Other Texts, Rare *Wing L 1908 7580.F.
392. Bedini, p.144.

## ACKNOWLEDGMENTS

This book would never have been researched, written, or published without Ed Kane and The Edward W. Kane and Martha J. Wallace Family Foundation. Ed's extremely generous and active support of previous comprehensive clockmaker books, authored by the scholar team of Donald L. Fennimore and Frank L. Hohmann III, carried forward to recruiting me for this Edward Duffield project. I was honored to undertake it and I sincerely hope that it meets his expectations and reaches the bar set by those superb books.

Many thanks, too, to Robert M. Hauser, APS Executive Officer. I am grateful to APS Interim Director of Publications, Peter Dougherty, for his valuable advice and guidance. The trio of anonymous reviewers provided frank and detailed comments which led to a much-revised and far-better treatment of Edward Duffield's life and work. I also am very grateful for the expert professional development editing provided by Thomas LeBien and Amanda Moon of Moon & Company. I am immensely grateful to the Incollect team of John Smiroldo, Phil Lajoie, and Marianne Litty for designing and producing a book that is far more beautiful than I could ever have imagined.

I am indebted to the late Ian M.G. Quimby (d.2002). In 1963, at the close of his two-year fellowship at Winterthur, he completed an unpublished biography and catalogue of Edward's work that remained the best source of related information until this project. He described fifteen Duffield clocks known at the time, he corresponded with several of their owners, and he hand-copied Edward's will and estate inventory. Those originals now cannot be located in Philadelphia's city records. His thesis title, "Edward Duffield: Artisan Gentleman," confirmed that Edward was far more than a leather-apron clockmaker.

Several current scholars, curators, and researchers were of great help to me. Philadelphia historians Torben Jenk and Joseph J. Menkevic generously provided me with many key materials, scans, and photocopies. Prior research by Winterthur emeritus curator Don Fennimore was invaluable.

Decorative Arts Curator Alexandra Kirtley of the Philadelphia Museum of Art, Baltimore Museum of Art Assistant Curator of Decorative Arts Brittany Luberda, Metropolitan Museum of Art Associate Curator of American Decorative Arts Alyce Englund, Chief Curator Karie Diethorn of Independence National Historical Park, and Colonial Williamsburg Foundation Curator of Furniture Tara Gleason Chicirda, each generously provided me with important information and viewings of Duffield clocks in their collections.

Historical Society of Pennsylvania researcher Margaret Maxey was prompt and thorough when answering my queries and in assisting Keirstyn Allulis, whom I recruited to research there and whom I also thank. Digital Archivist Andrew Williams was extremely helpful in providing the four HSP images in my text. Debbie Klak of the All Saints' Torresdale Episcopal Church showed me the Duffield family gravesite, and she shared with me her Duffield family genealogy and her careful transcription of Edward's worn gravestone.

Curators at libraries, societies, and smaller museums with Duffield clocks were equally giving of their time and information: Curator of Art and Artifacts Linda August at the Library Company of Philadelphia; Historic Annapolis former Curator of Collections Robin Matty Gower; Wyck Historic House Executive Director Kim Staub; Pottsgrove Manor Curator Amy Reis; Alexander Ramsey House Historic Site Manager Christine A. Jones; Free Library of Philadelphia Theatre Collection Curator Karin Suni; Joan Frankel of the Atlantic County Historical Society; and American Philosophical Society Associate Director for Collections and Exhibitions Mary Grace Wahl. Several private owners of lovely Duffield tall clocks kindly welcomed me into their homes. The Logan Family's mansion, Stenton, has no Duffield clock but Curator Laura Keim was generous with her scholarly expertise and deep knowledge of Philadelphia's colonial history.

Several professional photographers traveled with me to public and private owners of Duffield clocks to produce the fine photographs in the book's catalogue. They are individually credited there, and I thank them for their time, high-quality work, and obvious appreciation of the stately clocks they were documenting.

My longtime friend Robert C. Cheney, Executive Director and Curator of the Willard House & Clock Museum, created a stir with his April 2000, article in *The Magazine Antiques*. His compelling research about the English origins of early American clocks changed the landscape of horological history and guided my conclusions about clockmaking in eighteenth-century Philadelphia.

Special thanks are due to my good friend Jay Robert Stiefel. This eminent Philadelphia historian, author of a monumental APS-published book on cabinetmaker John Head, was instrumental from day one in generously and steadily providing me with key advice, contacts, published sources, encouragement, and special access that all contributed in large measure to the contents of this book. I am especially grateful for his providing the thoughtful Foreword. I also thank him for generously inviting me to admire, examine, and photograph his fine example of Duffield's work.

And finally, of course, deepest thanks and gratitude go to my wife of fifty years, Jeanne Schinto. A lifelong successful writer, whose skills far exceed mine and whose output of published books, articles, essays, and reviews dwarfs my own and most other writers, Jeanne remains my role model for how these kinds of years-long projects should be tirelessly crafted.
I would not be me, and this book would not exist, without her.

CATALOGUE OF CLOCKS & INSTRUMENTS

# Catalogue of Edward Duffield Clocks and Instruments

*Clock No.31, private collection, photo by Derek Dudek.*

This catalogue features seventy-one signed Edward Duffield clocks and instruments. The locations of forty-eight are noted, although private owners requested anonymity, and nearly all were professionally photographed. Those designated as "Location Unknown" are documented from older sources and photos. While every known Duffield clock and instrument is included, this cannot be a catalogue raisonné; more surely will come to light.

In the eighteenth century, "clock" almost always referred only to the movement and dial; the wooden cases crafted by joiners and cabinetmakers usually were identified separately. For example, in the 1752 estate inventory of Philadelphia's Casper Wistar, "Clock and Case" were noted in the parlor. Some early inventories did simply state "clock" as including the case, and today that distinction is entirely forgotten. The clock now is the complex machine combined with its furniture, from feet to finials.

All but one of the clocks in this catalogue are in tall wooden cases made in the Philadelphia area. (No.61 is a bracket/table clock clearly made in England.) Such pendulum-regulated weight-driven clocks, standing six to nine feet tall, were produced in Europe and North America for more than a half-century prior to Edward opening his shop in 1751.

Most people call them "grandfather" clocks. That term was unknown until the 1876 American Centennial celebration when Henry C. Work's song *Grandfather's Clock* became popular. Previously, "clock" or "eight-day clock" sufficed, and the maker was almost never named in estate listings. If inventoried clocks were dwarf (or grand*mother*), wall-hanging, or table models, they were identified in that way instead.

The clock sections of this catalogue are organized by case-bonnet styles. Although far from definitive and often overlapping in dates, this corresponds generally to chronology. First are cases with pagoda, sarcophagus, caddy, and round tops reflecting styles dominant in the early to mid-eighteenth century. More numerous are broken-arch scroll tops that follow—a Chippendale furniture style that surged in popularity in the 1760s. Plainer flat-top cases are next, although most are not currently located and some may be earlier and more appropriate among the first grouping. Sometimes a flat top suggests that an ornate top treatment has been lost, perhaps to accommodate lower room ceilings. For example, No.5 was purchased at auction as a flat-top, but had proper new top pieces recreated and added. Dwarf and bracket clocks, two uncased movements with dials, surveying compasses, and one sundial are in the final section. Not shown is a partial movement and badly worn dial recently donated to the American Philosophical Society by Edward's four-time great-granddaughter.

Details specific to each clock are provided to the extent available. Entries vary in length depending on whether clocks have been located, examined, and if their specifications are included in the cited references. Clocks in public institutions often are on display and benefit from curator-generated labels and online details. Clocks in private collections depend on specifics provided by their owners and sellers. Descriptions of unlocated clocks rely on viewable features and details published in the past.

Common characteristics of movements and dials are not needlessly repeated on every page. Readers should assume, unless otherwise noted, that each movement:
- was fabricated in the traditional eighteenth-century English manner from brass and steel components and is of the typical design;
- is weight-driven and of eight-day duration;
- has two trains of gears, one for timekeeping regulated by a recoil or deadbeat escapement and a seconds-beating pendulum, and the other with a rack-and-snail system for striking the hours on a top-mounted bell;
- has three appropriate steel hands;
- has an associated composite dial, made from brass and silvered brass, with Edward Duffield's engraved signature and with two holes for winding; and
- has a dial either square or arch-top with round applied signature boss or moon-phase indicator in the arch.

As noted in Chapter 3, side-view movement photos, typically published in auction catalogues and clock books, are minimally useful and provide few significant details. They simply verify that a proper standard movement is behind the dial. Such uninformative and repetitive photos are not included in the Catalogue, although readers are directed to Figures 3.7 and 3.8 for a sampling of them. If a side view is shown there, the Catalogue entry will note it as well as refer to other related movement and dial illustrations in Chapter 3.

A clock's provenance is offered when reported by present and past owners. As with nearly all antiques, provenance rarely is ironclad or complete for objects made long ago. It relies mostly on family and owner recollections and later chronologies that often cannot be independently confirmed. Writing and papers sometimes are found inside the cases. For example, Clock No. 23 has a yellowed paper tracing its lineage directly back to Edward Duffield, but the entries were written by someone seven steps removed from the clockmaker. Four more names, typed on newer white paper, conclude with the eleventh listed owner who perhaps did the typing.

Except for Clock No.3, which was commissioned by the APS in 1769, Edward's clocks are impossible to date precisely. Broad ranges based upon style and his years of clock work must suffice. His shop, first advertising foodstuffs, opened in 1751 and his identity as a horologist was not announced in print until five years later.

The 1752 estate inventory for prominent Philadelphian Caspar Wistar included an unnamed clock and case, with a value of £14, among the household goods. This was not the Duffield clock—Catalogue No.28—now standing at the Wyck Historic House, although they identify it as such. That assumption has been cited to imply that Edward was selling clocks at that early date. More likely, the inventoried clock was a 1730 Stretch clock known to have descended in the family and now in a private collection. Clock No.32 has "1752" roughly scratched on the back of its dial, but the case style is later and those numbers could have been inscribed by anybody at any time between then and now.

The Catalogue photographs show that no cases are identical, and that the cases vary in dimension, style, complexity, woods, and decorative elements. The cases are consistent in these features with most others made in Philadelphia that contain movements by Edward's contemporaries such as William Huston, David and Benjamin Rittenhouse, Jacob Godshalk, John Wood Sr. and Jr., Owen Biddle, Joseph Wills, and Peter and Thomas Stretch.

As discussed in Chapter Two, Edward did not make cases. Like his clockmaking colleagues in America and England, Edward housed his clocks in bespoke cases that were produced to client specifications by skilled woodworkers. Less ornamental cases may suggest initial purchase by local Quakers who oftentimes avoided displays of wealth. Case selections were made based on the owners' personal preferences and budgets, and on prevailing fashion.

Curators, dealers, and scholars of Philadelphia furniture have suggested plausible case-maker and carver attributions in several instances but none of these are firmly documented. Benjamin Randolph, George Claypoole, Henry Clifton, William Savery, Thomas Affleck, Jonathan Gostelowe, Nicholas Bernard, and Martin Jugiez are among the names often associated with the elegant cabinetry. Where case-maker attributions have been proposed, they are included along with the case's overall dimensions—in inches as overall height, width, and depth—and the principal wood of each clock. Cases are solid wood without veneers.

This author's scholarship in horology and history led to a focus upon Edward's life story, historical context, and eighteenth-century clockmaking. The book is not a treatise on Colonial Philadelphia furniture, so the Catalogue does not offer details about case aesthetics, designs, construction, carvings, turnings, inlays, moldings, glue blocks, secondary woods, hardware, etc. These are left to reader examinations of the images and to referenced sources including websites of museums and historical societies.

"DAPC" indicates that the clock is listed and described in the Winterthur Library Decorative Arts Photographic Collection Online Resource, a public-access database that has not recently been updated but remains a good source of helpful information. Case details are found in clock-related books and articles—many in the Bibliography— that mainly are by furniture-oriented authors who are not horologists. Like many of today's collectors and dealers, they target the clocks' wooden cases rather than the mechanical and metal components for which the clockmaker was directly responsible.

The existence of Duffield-signed instruments is not surprising. Like many colonial watchmakers and clockmakers, Edward had broad technical skills, which included engraving and dialing (the production of sundials), and fabricating surveying compasses otherwise known as circumferentors.

For a few thousand years sundials were the principal method of time-telling during daylight hours. These non-mechanical devices continued in regular use for setting mechanical clocks until the twentieth century. To show the correct local time, the inscribed lines must reflect the latitude of that locality. Edward engraved his sundial with his name, the year 1757, and "LAT.D. 40," which is his city's global north-south position. He also marked an outer ring with "equation of time" corrections, not commonly found on sundials. That table provided day-by-day conversions from the dial's readings that vary seasonally, compared to clock time that advances at the same rate throughout the year. Almanacs such as *Poor Richard's* always included a page of these tables.

Edward certainly engraved dies for two 1757 medals and for his signatures and city name on his brass clock dials, but possibly his sundial and compasses were made by others.

The complex engraved figures and patterns on the compasses are markedly different from one another, although he may have been demonstrating his singular engraving skills on each. Edward's surveying instruments are similar to examples produced by other American and English makers, who all needed specific knowledge and abilities to fabricate and engrave them correctly.

Such instruments were coming from England and could have been signed and retailed by Edward. A lengthy advertisement placed by importers Rivington and Brown in 1762 included surveying goods: "Theodolites, curcumferenters (sic), Gunter's chains, Hadley and David's quadrants, mathematical instruments, &ec."[384] A long advertisement in that same year was placed by "James Ham, Mathematical and Optical Instrument-maker," and listed dozens of products: "…Theodolites, Circumferenters, neatly silvered, with Ball and Sockets," along with many other instruments for surveying and "Brass Pocket Dials (sundials) fitted to the Latitude of Philadelphia."[385]

Another long advertisement placed in 1766 by instrument-maker Benjamin Condy on Front Street also offered a broad selection: "Hadley's and Davis's quadrants neatly finished in brass, ivory or wood, nocturnals, forestaffs, amplitude compasses, common ditto in brass and wood, crown and neat japanned hanging compasses, Gunter's scales and dividers, three and four feet telescopes with brass drawers and mahogany tubes, japanned ditto, theodolites, circumferenters…"[386]

Much earlier in 1740, when Edward was just ten years old, Charles Walpole already had advertised: "Mathematical Instrument Maker from London: Makes, sell and mends all sorts of instruments both for Sea and Land, at the Sign of the Davis Quadrant, at the Corner of McCoome's Alley, in Front-street, near Arch-street, Philadelphia."[387]

In Pennsylvania, the Chandlee-family clockmakers produced such instruments, as did Joel Bailey (1732–1797) of Chester County who worked with Owen Biddle to observe the 1769 transit of Venus and was elected to the APS the following year. Brothers Benjamin and David Rittenhouse made them; one example signed "Potts Rittenhouse"—Benjamin Rittenhouse and William Lukens Potts—is in the APS collection.[388] David devised a major improvement that applied a vernier scale to heighten accuracy."[389] Such vernier compasses became known as "Rittenhouse Compasses." A compass made by Elizabeth Town, New Jersey, clockmaker Aaron Miller (1710–1779) was sent in 1766 to Benjamin Franklin who presented it to London's Society for the Encouragement of Arts Manufactures and Commerce.[390]

A surveying compass signed (on the sighting arm, not its dial) by Benjamin Banneker (1731–1806), the free African-American naturalist, mathematician, astronomer and almanac author, sold at a Skinner auction in Massachusetts in 2020. Banneker in 1791 briefly joined the surveyors hired by Andrew Ellicott who were laying out Washington, D.C. and it is possible that he used this instrument during that work.

We may speculate that Edward learned surveying and its related instrument-making from tutors mentioned in Chapter Nine. These tasks also would have required him to have a firm understanding of higher mathematics, particularly geometry and trigonometry. He had access at the Library Company of Philadelphia to relevant technical texts, including William Leybourn's classic, first published in 1653, *The Compleat Surveyor*…[391] There are two references detailed in Chapter Three, within a ledger at the APS, indicating payments to Edward for work on surveying tools.

As an owner of extensive land holdings, Edward would have benefited, like many other landowners including George Washington, from undertaking surveying work and using the instruments himself to mark boundaries and lay out building locations. In 1769, Washington ordered from London an "18 Inch Circumferentor w. Sights to let down… £4.10.0,"[392] and he owned a full range of surveying instruments.

The Catalogue does not include unsigned clocks and instruments possibly attributed to Edward, although a few of these were discovered. Picturing a Duffield object in this Catalogue intends no guarantee of its authenticity, but each one appears substantially correct and appropriate. Definitive assessments of these clocks are best left to museum professionals, certified appraisers, and reputable sellers, when authenticity and valuations are being determined. No expert would assert that a Duffield clock today can be in precisely the same original condition in which he sold it two-and-a-half centuries ago.

While we hope and assume that more Duffield clocks will come to public attention, his clocks known to us today were cherished throughout their long histories. Other Duffield-signed clocks must have existed, but unfortunately were damaged beyond repair by fire or flood or insects, or they were simply discarded when no longer ticking or presentable.

With the surging availability of inexpensive smaller clocks manufactured in Connecticut beginning in the early 1800s, floor-standing clocks from earlier times became obsolete, old-fashioned, unreliable, and costly to maintain. Only when Colonial-Revival interest arrived at the time of the 1876 Centennial was there renewed appreciation for them that happily continues to the present day.

These stately clocks directly connect us with our Colonial ancestors who gazed upon them and who heard, as we do today, the same steadfast ticking and tolling. Edward's clocks, and those by his colleagues here and abroad, remain powerful symbols of traditional cultural values and of the stability, sophistication, and prosperity of people blessed to own them.

# Edward Duffield Clocks with Sarcophagus, Pagoda, Caddy, and Arched Tops

*Photo credits as indicated. Where not shown, courtesy of named auction houses or, if "DAPC", courtesy of Winterthur Museum, Garden & Library: Decorative Arts Photographic Collection.*

# 1
# POTTSGROVE MANOR

Pottstown, Pennsylvania

Mahogany case, 88.5" x 22.75" x 11.25"

1977.0890

Gift of Mrs. William H. Pomery (Edith Potts), Haverford, PA, direct descendant of John Potts (1710–1768) of Philadelphia. Dr. Benjamin Duffield, son of Edward, married John's daughter Rebecca in 1778 at the Manor but the clock's early provenance is unknown.

REFERENCES: DAPC. *The Mercury* (Pottstown, PA) October 24, 1969, p.19.

*Photos by John Wynn.*

## 2
## LIBRARY COMPANY OF PHILADELPHIA

Philadelphia, Pennsylvania

Walnut case, 99" x 23" x 12.25"

OBJ 865

Gift in 2003 by Richard Alan and Pamela Mones. Dr. Mones owns the LCP share formerly purchased in 1768 by Edward and held in his family until 1844.

REFERENCES: Stiefel pp.824–5.

*Photos by John Wynn*

## 3
## AMERICAN PHILOSOPHICAL SOCIETY

Philadelphia, Pennsylvania

Mahogany case, 82.5" x 23.5" x 10.25"

58.62

Unlike most clocks in the catalogue, this has a simpler sheet-brass dial and a single-train movement with deadbeat escapement. Not made for household use, it was commissioned by the American Philosophical Society in 1769 for timing the transit of Mercury. Edward in three weeks completed the movement that features a thick seatboard, thin movement plates, and top-mounted rate adjuster. In the Society's Transactions, John Ewing reported its use in Independence Square on November 9: "Still having the same instruments in our Observatory…. together with a new Time-Piece made by Mr. Duffield, of this city… we paid the utmost attention to the going of the clock, both before and after the transit." Edward was paid £15.17.6 when a committee deemed "the work good & the charge reasonable." Repaired 1925, 1927, and 1930 per APS curator reports.

REFERENCES: DAPC. Quimby 3. Multhauf p.49–50, figs. 20 & 22. Smith pp.2–3, fig.1. Rittenhouse & Ewing, p.82. Letter reports to the President and Council of the American Philosophical Society: 1926, 1928, 1931. *The Times*, Philadelphia, February 26, 1893, p.21.

*Photos by John Wynn*

# 4
## AMERICAN PHILOSOPHICAL SOCIETY

Philadelphia, Pennsylvania

Mahogany case, 82" x 16" x 9.5"

58.63

Like No.3, the movement of this clock has a single train of gears and does not strike the hours. Junius S. and Henry S. Morgan gifted the clock to APS in 1954. This may be the "Time piece in library" in Franklin's estate inventory and the clock Deborah Franklin in 1766 was reluctant to move while her husband was in London. The clock's shorter height could indicate that it was designed for a private domestic setting rather than a grander public space. The applied silver plates state that the clock passed from Franklin to his grandson Benjamin Franklin Bache (died 1798) and then to Bache's son Hartman (died 1872). However Multhauf cautioned: "The association of this clock with Franklin is not documented…" Hartman's son Richard Meade Bache inherited the clock and in his 1897 article, "The So-Called "Franklin Prayer Book," he wrote of "a clock of his own make, a gift to Franklin, is now in my possession." In 1907 the clock passed to his son Rene Bache. Banker J.P. "Jack" Morgan Jr. purchased the clock in 1933 and willed it to his two sons named above. This is one of four by Edward with a rare spherical moon in its dial arch. Franklin while in England perhaps saw such a clock by Thomas Ogden, likely originator of this feature, and requested it from Edward. Displayed in the Benjamin Franklin Tercentenary Exhibition and in the associated book.

REFERENCES: DAPC. Quimby 3. Multhauf, pp.50–51. Eckhardt p.285. Smith pp.16–17, fig.9. Eckhardt 1959, p.132. Talbot p.137, Bache p.7. Figure 3.12.

*Photos by John Wynn*

## 5
## PRIVATE COLLECTION

New Jersey

Maple case, 108" x 22" x 12.75"

Sold at a Pook & Pook auction. Descended in the Griscom family from James Cooper Griscom who was related to the Coopers of Camden, NJ, on Martins Pike near Woodbury. The case when sold was flat-top but a new appropriate stepped top and three finials were crafted by Alan Andersen.

REFERENCES: DAPC. Pook & Pook September 28-29, 2007, Lot 779. *Maine Antique Digest,* December 2007, p.3-C. "Ancestral Home of Griscoms, Members of 8th Generation," *Courier-Post,* Camden, NJ, p.3.

*Photos by John Wynn*

# 6
# PRIVATE COLLECTION

New Jersey

Walnut case, 101"

Purchased in 2012 at Pook & Pook with stated provenance from "A New Jersey Educational Institution." This was the Camden County Historical Society that informed the current owner that the clock "came to us in 1986 from the Albertson family of Camden County…. we don't know when the clock came to them, or under what circumstances." In 1682 William Albertson acquired land in West Jersey that in 1709 went to Josiah Albertson. In 1743 he built a large brick house where the clock may originally have stood.

REFERENCES: Pook & Pook, January 14, 2012, Lot 839. Figure 3.8.

*Photos by John Wynn*

# 7
## PRIVATE COLLECTION

Pennsylvania

Walnut case, 100" x 21" x 11"

Sold to its current owners by dealer Philip H. Bradley at a January 30, 2016, antique show in York, Pennsylvania. The sales receipt provided no provenance but stated: "An early clock by Edward Duffield of Phila in its original case c1750, walnut in sarcophagus top, 1/4 columns in waist, attached hood columns, architectural molded feet. Sarcophagus above frieze restored, some repair to original feet. Arched Brass Dial with engraved Boss. Case in old finish, Case outside Phila."

REFERENCES: "166th Original Semi-Annual York Antiques Show and Sale", Lita Solis-Cohen, *Maine Antique Digest,* February, 2016.

*Photos by Nathan Merkel*

# 8
# FREE LIBRARY OF PHILADELPHIA

Parkway Central Library

Mahogany case, 103" x 23" x 12"

Dr. Bushrod Washington James (1836–1903) gifted this clock in 1903. Papers inside the case state: "This clock was among the household goods given to Miss Henrietta Potts on her marriage to Isaac James on March 26th, 1801. It belonged to her father Capt. Potts or to her grandmother Mrs. Rebecca Grace. It was bequeathed by Isaac James to his eldest daughter in these words: "Second. I give and bequeath to my daughter Anna Potts James my eight day clock," Date of will, 29th July 1864, admitted to Probate, Feby. 2d. 1874. And bequeathed to her nephew in these words: "I give and bequeath the old tall eight day mahogany case clock to my nephew Doctor Bushrod W. James when my sister Martha and brother Samuel may have no further use for it." Date of will, Jany. 18th, 1879, Admitted to Probate Nov. 6th, 1879."

REFERENCES: Provenance papers inside clock case. Figure 3.1.

*Bushrod Washington James*

*Photos by John Wynn*

Photos by John Wynn

# 9
# PRIVATE COLLECTION

New Jersey

Walnut case, 97"

According to the present owner, this has been in her family for several generations. They are descendants of John Chambers who reportedly bought the clock from Edward. A handwritten Chambers genealogy inside the clock extends back to an Irish immigrant in 1726. A card affixed to the backboard reads,

"Adolph Yunker, 157 E. Front Str. Trenton, Clnd, one cord gut right, August 23, 1907." June 14, 1965, receipt inside the clock shows a charge of seven dollars to clean, oil, and fit a new second hand.

## 10
## COLUMBIA UNIVERSITY

New York, New York

Walnut case, 110.5" x 23" x 12"

One of four Edward movements with a rare spherical moon in its dial arch. The top piece and finial are later additions; height without them is 96". Two dowel holes in the flat top of the bonnet could indicate that there originally was a different decorative top piece.

REFERENCES: Figures 3.1 and 3.12.

*Author photos*

# 11
# PRIVATE COLLECTION

Pennsylvania, Walnut case, 109" x 22.25" x 11"

Sold at Brunk Auctions, Asheville, NC, from the collection of the late dealer C.L. Prickett. Restoration including appropriate new feet and reconstructed pagoda top by Alan Andersen who states that the finish likely is original.

REFERENCES: Brunk Auctions, September 16–27, 2022, Lot 1094.

*Photos by Nick Gould*

## 12
## LOCATION UNKNOWN

Walnut case, 95" x 20.25" x 11"

David Stockwell illustrated this in *The Magazine Antiques,* January, 1969, and again in March, 1991, stating "The early case is probably by Nathanial Dowdney, Philadelphia cabinetmaker." Offered at a Wiederseim Associates auction November 27, 2010.

REFERENCES: DAPC. *The Magazine Antiques,* January, 1969 and March 1991. 1973 Delaware Antiques Show program, p.62.

## 13
## LOCATION UNKNOWN

Cherry case, 85"

Sold at Wiederseim Associates auction, Chester Springs, Pennsylvania, April 21, 2011, Lot 279.

Catalogue No.12, *The Magazine Antiques*, January 1969, p.1.

# Edward Duffield Clocks with Broken-Arch and Scroll Tops

*Photo credits as indicated. Where not shown, courtesy of named auction houses or, if "DAPC", courtesy of Winterthur Museum, Garden & Library: Decorative Arts Photographic Collection.*

# 14
# ATLANTIC COUNTY HISTORICAL SOCIETY

Somers Point, New Jersey

Walnut case, 103" x 23.5" x 11.5"

#95.11.1

Gift July 26, 1995, of Mr. David Jones Somers who claimed direct descent from David Jones (1740–1785), cordwainer of Philadelphia (will proved February 7, 1785), who reportedly was the clock's original owner. The family genealogy is in the society's object file, along with a 2013 assessment of the clock's features and condition by a local clockmaker. Ian Quimby noted "Collection of Mr. Hubert Somers" and "simple molded base… it is unusually short and was probably cut down."

REFERENCES: DAPC. Quimby 13. Sperling. Eckhardt p.285 fig.4. Society website. Figure 3.7.

Photos by John Wynn

*Photos courtesy of BMA*

## 15
## THE BALTIMORE MUSEUM OF ART

Baltimore, Maryland

Mahogany case, 103" x 22" x 11.75"

BMA 1960.41.19

Gifted by Phillip B. Perlman of Baltimore, Secretary of State of Maryland (1920–1923), Baltimore City Solicitor (1923–1927, 1931–1943), and Solicitor General of the United States (1947–1952). He personally argued the landmark anti-discrimination case of Shelley v. Kraemer. The engraved silvered-brass sheet dial was uncommon for Edward and less costly than the composite dials he mostly favored. Three 19th-century repairer names and dates are boldly chalked on the inside of the backboard. A printed repair sticker inscribed to Dr. Newbold, dated "9-16-13," was affixed to the movement seatboard by "William Gibbons, Clock Maker, 5 South 40th Street, Philadelphia."

REFERENCES: DAPC. Quimby 11. Sperling p.591. *The Magazine Antiques*, November 1961, p.474. Elder p.115. Museum website and its "Collecting" article by Kerr Houston, March 29, 2023.

# 16
## PRIVATE COLLECTION

Virginia

Mahogany case, 110" x 23" x 11"

In 1932 this was the property of Mrs. Thomas A. Curran. Quimby later stated: "owned by Robert B. Haines, III…. but no information or photograph available." In June, 1996, it was sold at Christie's with case carving attributed to Philadelphia carvers Nicholas Bernard and Martin Jugiez. In May, 2004, Philip Bradley Company offered it as from: "The Estate of James Curran." In 2019 it was sold to the present owner. This style of pitched-pediment broken-arch top is unique for Edward's clocks but seen in English furniture.

REFERENCES: DAPC. *Antiquarian* April 1930 (per DAPC). *The Magazine Antiques,* April, 1930, ad for Estate of James Curran. Stretch p.235. Quimby 15. Christies New York, June, 1996, Lot 156. *The Magazine Antiques,* May 2004 ad by Philip H. Bradley.

*Photos by Gregg Vicik*

# 17
# PRIVATE COLLECTION

Pennsylvania

Walnut case, 119" x 24.5" x 13.5"

Known as "the Wright Family Clock," it was never out of the family until sold, on November 21, 2010, at Freeman's auction in Philadelphia, to dealer James Kilvington and then to the present owner. Freeman's stated:

*Merino Hill House, Wright family homestead, Monmouth County, New Jersey*

"ownership can be traced to Samuel Gardiner Wright (1781–1845), a highly successful Quaker Philadelphia merchant, investor, iron master, [and] New Jersey gentleman farmer .... The tall case clock stood for 150 years in the Merino Hill House." The property descended from David Wright (d. 1791), whose probate inventory listed a highly valued "Clock." Repair card by the late 19th-century clockmaker Ben[ajah] Budd of Mount Holly, NJ. The clock has one of four dials by Duffield with a rare spherical lunar indicator.

REFERENCES: Freeman's November 21, 2010, Lot 557. Lita Solis-Cohen, "American Furniture and Decorative Arts and the Fifth Pennsylvania Sale," *Maine Antique Digest,* February, 2011. Probate inventory of David Wright, July 19, 1791. Figures 3.7, 3.8, and 3.12.

*Photos by John Wynn*

## 18
## WINTERTHUR MUSEUM, GARDEN AND LIBRARY

Winterthur, Delaware

Mahogany case, 122.5" x 24.25" x 12"

1952.0247

Gift of Henry Francis du Pont. Case restored with new center cartouche and upper feet. Regarding the deadbeat-escapement movement, Hohmann noted: "The brass plates are somewhat thinner than those of most Pennsylvania movements, and the wheels, arbors, and pinions are diminutive." LaFond noted a T-shaped front movement plate and decorative scoring on movement gears and pillars, carved decoration on the scroll board that probably was applied later since he found it crowded and of lower quality, and a moon disk painted with moons, stars, and sky.

REFERENCES: DAPC. Quimby 10. LaFond (unpublished manuscript). Hummel 1976 p.129 Fig.120. Downs No.204. Eckhardt Fig.5. Hohmann No.46. Shaffer p.27. Distin Fig.60. Thomson Fig.18. Krill p.117, pp.120–121. Winterthur Online Collections.

*Photos courtesy of Winterthur Museum, Garden & Library*

## 19
## COLONIAL WILLIAMSBURG

Williamsburg, Virginia

Mahogany case, 113" x 23.5" x 11.5"

1930.583

Sold at the Howard Reifsnyder famous auction in 1929. Quimby caption: "Tall clock in the Governor's Palace at Colonial Williamsburg." Cleaning dates on movement date back to William Reeves in 1783 and Samuel Norton in 1798 in Philadelphia, then up to Adolph Yunker in 1888. The movement has a deadbeat escapement. One of four known Edward dials with a rare spherical moon in its arch.

REFERENCES: DAPC. Quimby 9. Distin fig.61. Winchester p.75. Robey p.853. Cooper pp.6–7. 1929 Girl Scouts Exhibition p.606. Pennsylvania Museum Bulletin May, 1924, p.163, plate 5. "Exhibition of Furniture of the Chippendale Style," Pennsylvania Museum of Art, Philadelphia, 1924. Hohmann No.47. Eckhardt p.286, Fig.6. Anderson Galleries American Art Assoc. Reifsnyder Sale, April 24–27, 1929, Fig.667. Redwood Board Room, Carpenter Collection. John D. Rockefeller, Jr., Christies, June 19, 1996, Lot 156. Museum website. Figure 3.12.

## 20
## AMERICAN PHILOSOPHICAL SOCIETY

Philadelphia, Pennsylvania

Walnut case, 104" x 21.5" x 11"

1966.0036

Ian Quimby recorded Henrietta Bache Jayne of Wallingford as the former owner. It may be in Franklin's estate inventory as the unnamed clock on the stairs. Descended through the Bache family. This or No.4 may be the clock mentioned by Richard Meade Bache (1830–1907), great-great-grandson of Benjamin, as a gift from Edward to Franklin. Per Smith, Benjamin Randolph may have made the case.

REFERENCES: DAPC. Quimby 2. Eckhardt fig.2. Smith p.17–18. *The Magazine Antiques,* October, 1972, p.683. The Benjamin Franklin Tercentenary Frankliana Database. Figure 3.8.

*Photos by John Wynn*

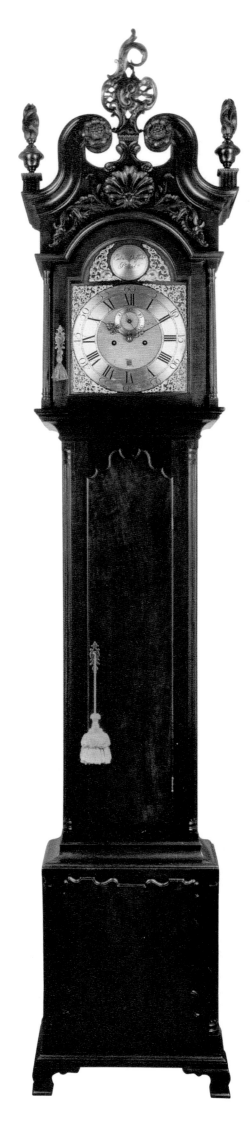

## 21
## PRIVATE COLLECTION

New York

Mahogany case, 100" x 22.5" x 11.25"

Provenance was noted by Sotheby's in 2005: Samuel Howell, Jr. (1748-1802), Samuel Emlen Howell (1772-1839), William Rumfort Howell, Lardner Howell, Mrs. John Randolf Young of Whitford, PA, to consignor. 1798 wedding present to Samuel Emlen Howell who married Mary Whitlock Dawes (1778-1846). It may originally have been owned by Samuel's father, a Philadelphia ship captain and Revolutionary War veteran. His son followed him as a prominent merchant in the China Trade. The case is attributed to Henry Cliffton and Thomas Carteret. A 1966 repair sticker from William A. Heine, Philadelphia, named customer Howell. Sold at Sotheby's to dealer Todd Prickett for a client.

REFERENCES: DAPC. Quimby 12. Sotheby's May 19, 2005 Lot 251. Hohmann No. 45.

*Photos courtesy of owner*

# 22
# METROPOLITAN MUSEUM OF ART

New York, New York

Mahogany case, 105" x 19.5" x 10.5"

2014.223

Purchased in 2014 by the museum's William Cullen Bryant Fellows and Close Friends Gifts in honor of the retirement of curator Morrison H. Heckscher. It was owned by three generations of a Bethesda, Maryland, family from the early twentieth century until 2010 when it was purchased at Christie's by a West Coast collector. Such cockerels are seen atop tall clocks with movements by Frederick Maus of Philadelphia. The rooster symbolized Apostle Peter, the patron saint of clock- and watchmakers.

REFERENCES: Christie's, September 29, 2010, Lot 117. Hohmann, pp.331–332. Chris Storb email to author after examining case. Milley, p.130. Mayr, p.7.

*Photos courtesy of Metropolitan Museum of Art*

# 23
# PRIVATE COLLECTION

Minnesota

Walnut case, 101" x 22.5" x 11.25"

Dealer Gary Sullivan offered this for sale in 2014 and it was sold to the present owner. Made for the Johnson family who were Quaker abolitionists with a large home, in the Germantown area of Philadelphia, that now is a national historic landmark and Underground Railroad site. A February 16, 1775, issue of the *Massachusetts Spy* was pasted inside the backboard to cover a crack. The dial is signed with Edward's full name, not the more typical Edw. or E.

*1768 Johnson House, Germantown, Philadelphia*

REFERENCES: Website of Gary Sullivan Antiques. Johnson House Historic Site.

*Photos by Dan Denehy*

Photos by Matt Buckley

## 24
## PRIVATE COLLECTION

Rhode Island

Walnut case, 106"

Sold at Freeman's auction on November 14, 2013, then in 2020 at Beaver Creek Antique Market in Hagerstown, Maryland, to dealer Gary Sullivan and then to the present owner. The author completely disassembled, overhauled, and photographed all of the movement and dial components including the "T-shape" brass front plate needed to mount the dial with its large revolving moon-phase indicator. These front-plate extensions are seen on movements by several makers from the period. Repair sticker from William A. Heine, Philadelphia, recording work for Mr. W.L. Coates, Jr. in Wilmington, Delaware, March 9, 1955.

REFERENCES: Freeman's November 14, 2013, Lot 432.

## 25
## PRIVATE COLLECTION

New Jersey

Walnut case, 108.25" x 23.5" x 12"

Sold at Alex Cooper Auctioneers, Inc., Baltimore, Maryland, June 16, 1980, and then at Sotheby's in 1995. In 1932 it was displayed in the Harlem Lodge office of Dr. William Rush Dunton (d. December 1966) who inherited the clock from his father. Brooks Palmer 1952 letter stated that it was a "splendid example". Charles Robert Lynch of Reistertown, Maryland, bought it from the Dunton estate. Sotheby's photo showed a center flame-form finial, not the later cartouche carved by Alan Andersen.

REFERENCES: DAPC. Sotheby's June 22, 1995, Lot 262. *The Catonsville Herald and Baltimore Countian,* Friday, November 4, 1932. Letter dated January 22, 1952, to Dr. Dunton from Brooks Palmer.

*Photos by John Wynn*

Photos by Nathan Merkel

## 26
## PRIVATE COLLECTION

Pennsylvania

Walnut case, 100.5"

Sold in 2009 at Pook & Pook, Downington, Pennsylvania, to the current owner. Edward's full name, not the typical Edw. or E. is engraved on the dial.

REFERENCES: DAPC. Pook & Pook, October 2–3, 2009, Lot 350.

*Photos by Dan Denehy*

## 27
## ALEXANDER RAMSEY HOUSE

St. Paul, Minnesota

Mahogany case, 92" x 20" x 10.75"

1988.155.700

Gifted to the Ramsey House by Margaret B. Witmer, a great-great niece of Mr. Ramsey, in 1979. Previously willed to Eliot F. Porter, Jr., son of Eliot F. Porter (1901–1990) of Santa Fe, New Mexico, who was awarded a Guggenheim Fellowship in 1941 for his bird photography.

*Eliot F. Porter, Jr.*

REFERENCES: DAPC. Minnesota Historical Society report for object 1988.155.700.

## 28
## WYCK HOUSE

Philadelphia, Pennsylvania

Walnut case, 99" x 21.75" x 10.75"

79.123

Reportedly owned by Casper Wistar (1696–1752) as the unnamed "Clock and Case" listed in his estate inventory. However, Edward was just 22 years and not yet selling clocks, so the inventory likely referred to a different clock. Reuben Haines, Casper's grandson, certainly owned this clock in 1812 per a repair bill, and in 1820 Reuben Haines III moved his family to Wyck full-time, perhaps with this clock. A reception for Lafayette at Wyck House on July 20, 1825, placed him with the clock. Appraised in the 1912 estate of Jane Reuben Haines, seventh generation owner of the house, who had the clock bolted to the wall to prevent its removal. Repairs were made in 1922 and 1930 per the diary of Casper Wistar Haines. Carolyn Wood Stretch viewed it on May 9, 1931, in preparation for her 1932 article on Colonial Philadelphia clockmakers.

REFERENCES: DAPC. Casper Wistar estate inventory April 4, 1752. Moss p.135. The Wyck Association Collection Catalogue. Boudreau p.299. Stretch pp.226-227. *The Magazine Antiques,* August, 1983, p.282.

*Photos by John Wynn*

*Photos by John Wynn*

## 29
## HISTORIC WAYNESBOROUGH

Paoli, Pennsylvania

Walnut Case, 97" x 22.75" x 11"

WL07.06.01

Donated in 2007 to the Anthony Wayne Foundation by Dr. Richard A. and Pamela R. Mones on recommendation of dealer H.L. (Skip) Chalfant. Replaced finials and feet are noted. Presently on loan to Historic Waynesborough.

REFERENCE: Historic Waynesborough catalog information sheet dated November 26, 2007.

# 30
## WORCESTER ART MUSEUM

Worcester, Massachusetts

Mahogany case, 105.5"

1946.20

1946 bequest of Judge John M. Woolsey (1877–1945), New York, NY, a Walpole Society member and the U.S. District Judge who in 1933 ruled that James Joyce's *Ulysses* be admitted into the United States. Serviced in 1970 by Bradford W. Cheney and in 1984 by his son Robert C. Cheney.

REFERENCE: Reutlinger p.58, Fig. 69.

*Judge John M. Woolsey*

*Photos courtesy of Worcester Art Museum*

# 31
# PRIVATE COLLECTION

Connecticut

Walnut, 108" x 21" x 10"

Purchased in the 1990s from the late Edward F. LaFond, Jr., a clock dealer, restorer, and scholar in Mechanicsburg, Pennsylvania. The movement and period case may have been associated later.

*Photos by Derek Dudek*

Photos by John Wynn

## 32
## INDEPENDENCE NATIONAL HISTORICAL PARK

Philadelphia, Pennsylvania

Walnut case, 92.5" x 20.5" x 10.5"

INDE-02913

April 28, 1975, gift of Mrs. William R. (Anna Ruth) Purcell, Doylestown, from a family of early settlers of Bucks County. Early repair names and dates are chalked inside the case. Restored in July, 1776, by Robert Whitley of Solebury, Pennsylvania. "1752" is roughly scratched on the back of the dial but this is an unlikely early date for the clock.

REFERENCE: June 15, 1988, Museum Catalog Record - Cultural Resources.

Photos by Greg Staley

## 33
## PRIVATE COLLECTION

Virginia

Mahogany case, 94.5" x 20.25" x 10.75"

According to the present owner, this has been in his family since the 1800s. It is referenced in the 1919 will of his great-great-grandfather E. Burgess Warren, a Philadelphia builder, roofer, and collector of paintings, who listed it as the old family clock on the stairs.

## 34
## PRIVATE COLLECTION

Louisiana

Walnut case, 101" x 20" x 11"

According to the present owner, this has been in his family since the 1840s and was in Baltimore until around 1947.

*Photos by Seth Boonchai*

## 35
## PRIVATE COLLECTION

North Carolina

Walnut case, 99.5" x 21.5" x 11.5"

The present owner has had this for about twenty-five years and it descended in her family as far back as 1901. There is evidence that the top crest and finials were shortened and feet were removed, possibly to stand in a room with lower ceilings. Boldly chalked inside the case are several repair names and dates: "July 1, 1819, John Pittman; May 24, 1823; Widenhofer, Oct. 27, 1875 (also inside waist door): John Kirschnek Aug. 12, 1905, Jan. 1, 1915, Feb. 19, 1916" who was a local jeweler in Media, Pennsylvania. He sold his business at auction in 1928, and died at age 73 in 1934. Another Media repairman, A.R. Farreny, serviced the clock in March, 1951, per his card tacked on the seatboard.

 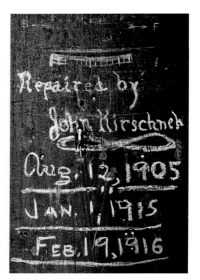

*Photos by Brandon Scott*

## 36
## PRIVATE COLLECTION

California

The owner states that this has been in his family for centuries. He grew up in Connecticut and his family came to New England in the 1660's.

*Photo courtesy of owner*

# 37
## LOCATION UNKNOWN

Walnut case, 101" x 22.5" x 10.5"

Sold by Skinner Inc., Marlborough, Massachusetts, in 2022 from the collection of Cy Felheimer of New Jersey. A small photo of its bonnet and dial was shown in Whisker, courtesy of the former owner. Felheimer died in 2006 and his obituary noted his serious clock collecting and active membership in the NAWCC including a Fellow Award and presidency of Chapter 1.

REFERENCES: Whisker p.164. Skinner March 31, 2022, Lot 1276.

*Photos courtesy Bonhams Skinner*

*Photos by John Wynn*

## 38
## WILLIAM PACA HOUSE

Annapolis, Maryland

Mahogany case, 101" x 19.5" x 9.25"

Advertised by David Stockwell in *The Magazine Antiques,* September, 1937. Donated to Historic Annapolis in 1977 in memory of Edward Boetler Passano (d.1946), owner of Waverly Press, by his son who claimed that the clock stood in their office until loaned to Historic Annapolis in 1975. There is no known connection of the clock to William Paca, an 18th-century University of Pennsylvania graduate, Philadelphia lawyer, and signer of the Declaration of Independence. The simpler lower-cost silvered-brass sheet dial is unusual for Duffield clocks.

REFERENCES: DAPC. *The Magazine Antiques,* September, 1937, inside back cover.

# 39
## AUTHOR'S COLLECTION

Andover, Massachusetts

Walnut case, 97" x 21.5" x 11"

Papers inside the case show the clock purchased by T. (Thomas) Williams, Jr. in July 1878, repaired by William A. Heine in 1949 for David Williams and again in 1962 for Mrs. Brinton (Mary Williams Brinton) whose 1972 autobiography, *Their Lives and Mine,* includes references to the clock. The Brinton home at 2008 Delancey Street in Philadelphia now is the location of the Rosenbach Museum. Former collection of Peter A. Pfafffenroth, then to Pennsylvania antiques dealer Kelly Kinzle, then to the author in August, 2023. An early replacement bell with cast-in name of John Wilbank (1788–1843), Caster, Philadelphia, who in 1828 provided a replacement for the damaged Liberty Bell. He was responsible for saving that iconic American relic from being melted as part of his compensation. The three finials are replacements carved by Alan Andersen.

REFERENCES: Figure 3.3 (disassembled movement)

*Mary Williams Brinton at her wedding in 1936.*

*Author photos.*

## 40
## LOCATION UNKNOWN

Pictured in Whisker book as courtesy of Pearl and Ira Prilik. Ira reported in 2020 that he possibly recalls selling the clock to a California collector in 1992.

REFERENCES: Whisker p.164 Figure 19. Email with author.

## 41
## LOCATION UNKNOWN

Walnut case, 99" x 19.5" x 10"

Per Sotheby's January 1992 catalogue, "The Lunette painted with a Coat of Arms with the inscription by the name of Leech Moore." Also offered at Sloans auction in July, 1989. Although probably unrelated, there was a Hannah Leech who married Jacob Duffield, a cousin of Edward, in 1742 at Christ Church. They had several children.

REFERENCES: DAPC. Sloans July 7-9, 1989, Lot 2499. Sotheby's January 23-25, 1992, Sale #6269, Lot 1073, unsold.

## 42
## LOCATION UNKNOWN

Walnut case, 101.5" x 18.5" x 10.5"

The clock was sold without provenance or documentation by Israel Sack to Dr. Bruce Berryhill in 1975, then returned to Sack some years later for resale. When first marketing the clock, Sack had erroneously published another clock's provenance as this one's, an error repeated by Dworetsky.

That provenance instead pertains to Catalogue Number 17, a meticulously documented clock which had never left the family until sold at auction in 2010.

REFERENCES: DAPC. Dworetsky p.3. *Israel Sack Collection,* Vol.3, p.640. DAPC.

## 43
## LOCATION UNKNOWN

Maple case, 103"

Pook & Pook auction catalogue stated: "Finials replaced. Patches at hinges… Feet replaced… Estate of Marie Schwarz (d. 2013 at age 93), Philadelphia, Pennsylvania." Per report by Lita Solis-Cohen: "The Saturday session opened with furniture, decorations, and jewelry from the estate of Marie Schwarz of the Schwarz Gallery in Philadelphia, whose apartment at the Barclay was often open for tours of visiting groups…. The buyer from Ohio in the salesroom seemed pleased to get it."

*Marie Devlin Schwarz*

REFERENCES: DAPC. Distin fig.170. Bailey p.25, fig.19. Pook & Pook April 25, 2014, lot 527. *Maine Antique Digest,* January 2015.

*Courtesy of Chris Bailey*

## 44
## LOCATION UNKNOWN

Mahogany case, 93.5" x 22" x 11.5"

Was in possession of Beaveau Borie III. Per email from Peter Nalle: the first Beauveau Borie (1846–1930) married Patty Duffield Neill (1846–1940), great-granddaughter of Edward Duffield. A flat-top clock also descended in the Borie family. According to Palmer, p.184, "B. Borie, Whitemarsh, Pa. owns two, one 5' high." Quimby catalogue stated: "Lunette has sunburst painted in blue and gold with a face…. deadbeat escapement…. now disassembled…. It may have been one of those mentioned in Duffield's inventory." October 24, 1962 letter to Ian Quimby from Henry Peter Borie stated: "My family is directly descended from Edward Duffield through my grandmother, the late Mrs. Beauveau Borie, who was Patty Duffield Neill. I have the grandfather clock which appears in the Antiques article by Mr. Eckhardt. My brother Beaveau Borie of 318 W. Springfield Ave. has another grandfather (this example), but unfortunately the case was badly damaged when it fell from its wall support."

*Beauveau Borie (1846–1930)*

REFERENCES: DAPC. Quimby 7. October 24, 1962, letter to Ian Quimby from Henry Borie reporting that clock was in pieces. (No image of complete case.)

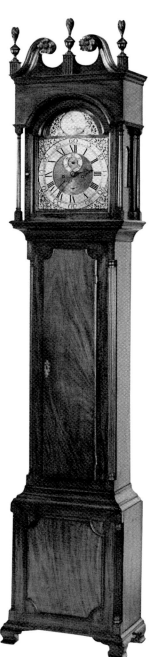

## 45
## LOCATION UNKNOWN

Mahogany case, 100.25"

Per ad by C.L. Prickett: "Provenance: Thomas Evans, original owner, married to Hannah Morris; to Susanna Evans Anderson, daughter, married Alexander Anderson; to Samuel V. Anderson, son, married Sarah Hayes Wickersham; to William V. Anderson, son, married Annia Penrose; to Emma Anderson Cabeen, daughter, married Robert B. Cabeen; to Francis von Albade Cabeen, son, married Mary Lea Cabeen; to Francis von Albade Cabeen, Jr., son, married Anne Amlen Hamersley Cabeen; to Edmond H Cabeen, son; to M. Lea Cabeen."

REFERENCES: *The Magazine Antiques,* December 1, 1990, p.1129. C.L. Prickett Brochure #2, p. 21.

## 46
## LOCATION UNKNOWN

Walnut case, 94"

Pictured with caption in Distin book. DAPC text stated: "The boss in the lunette may be inscribed with the maker's name, but this is unclear in the photo."

REFERENCES: DAPC. Distin p.29, fig.42.

## 47
## LOCATION UNKNOWN

Walnut case, 93.75" x 23.25" x 11"

Noted from ex-collection of Mr. & Mrs. Harold L. Murray, Jr. of Radnor, Pennsylvania. Footnote in a 2009 Chipstone essay mentioned an antique box from the "collection of Harold L. Murray, Jr. of West Chester, Pa." The case appears unusually wide, more typical of later cases made in the north of England.

REFERENCES: DAPC.

## 48
## LOCATION UNKNOWN

Walnut case, 87.5" x 17.5"

Formerly owned by Mr. & Mrs. Douglas Oliver of Philadelphia. A 1976 article identified Mr. Oliver as vice president of Girard Bank and later of Franklin & Marshall College. His Germantown home (but not this clock) was pictured on page 156 of *American Country Houses of Today* by Alfred Hopkins, 1927.

REFERENCES: DAPC. *Philadelphia Daily News,* Philadelphia, Pennsylvania, March 24, 1976, p.2.

*1927 view of Oliver home*

## 49
## LOCATION UNKNOWN

Mahogany case, 93.5" x 22" x 11.5"

Owned by Mrs. James B. White of New Castle, Delaware. Per Ian Quimby: "The clock was previously owned by John D. Cannon (c.1800–c.1895), Mrs. White's grandfather, who gave it to Mrs. White's mother, Elizabeth Rockett. Cannon is thought to have received it as a gift from a lady who left Philadelphia to go to England."

REFERENCES: DAPC. Quimby 8. Eckhardt p.285. Palmer p.184. October 25, 1962 letter from Ian Quimby to Mrs. White; February 1, 1963 reply from Mrs. (Emily) White.

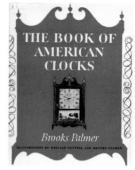

## 50
## LOCATION UNKNOWN

Mahogany case, 97.5"

Sold in 1972 at auction by Sotheby Parke Bernet in Los Angeles. The catalogue noted: "Repaired several times between 1817 and 1912 as noted on the inside of the case."

REFERENCES: Sotheby Parke Bernet Sale #3438, November 16–18, 1972, Lot 676.

*(facing page, dial photo only)*

## 51
## LOCATION UNKNOWN

Unknown wood case

Offered in 2010 for sale to an antiques dealer who received photos. Clock was not bought due to missing scrolls and feet. More recently, the former owner could not recall who bought the clock subsequently.

REFERENCES: Information and photo courtesy of anonymous dealer.

## 52
## LOCATION UNKNOWN

Walnut Case, 100"

Offered but not sold at May 12, 2016 Brunk auction in Asheville, North Carolina. From the collection of Dr. J.C. Gilbert, Fort Lauderdale, FL. Auction listing stated: "saddle board replaced (works probably a good match to case)."

REFERENCES: Brunk Auctions, May 12, 2016, Lot 475. Obituary of Dr. Gilbert published in *Sun-Sentinel* on March 27, 2013.

*Courtesy of Brunk Auctions, Asheville, North Carolina.*

# Edward Duffield Clocks with Flat Tops, Dwarf Clock, Bracket Clock

*Photo credits as indicated. Where not shown, courtesy of named auction houses or, if "DAPC", courtesy of Winterthur Museum, Garden & Library: Decorative Arts Photographic Collection.*

# 53
# PRIVATE COLLECTION

Pennsylvania, Pine Case, 89.5" x 19.5" x 14.5"

This clock has a square dial, single hand, and thirty-hour brass-plate pull-up movement with countwheel striking, perhaps one of the first clocks made by Edward. It is the only example that is not eight-day and without rack-and-snail striking. It formerly was owned by Clyde Fahrney of Waynesboro, Pennsylvania. Ian Quimby referred to it as "crude" and "old-fashioned." Eckhardt suggested that the case, "with its three boxlike sections, has a country look about it; the use of fielded panels and large H hinges almost suggests the work of a house carpenter." The presence of new wood supporting the old seatboard likely indicates that the case and movement were joined later. The plain pine case was painted in folk-art style by the current owner's restorer, the late Kendl Monn. The clock was offered in January, 2003, at Sotheby's; again at Sotheby's in January, 2004; and again at Pook & Pook in October, 2004 where it sold to the current owner. Although Sotheby's 2004 catalogue description noted "cheek apparatus replaced," recent movement photos by the author show no indication of any modifications to a standard anchor-escapement long-pendulum movement.

REFERENCES: DAPC. Quimby 1. Eckhardt Fig.1. Sotheby's, January 15, 2003, Sale #7865, Lot 542. Sotheby's, January 15, 2004. Sale #7959, Lot 404. Pook & Pook, October 8, 2004, Lot 697.

*Author photos*

*Photos courtesy of Philadelphia Museum of Art.*

## 54
## PHILADELPHIA MUSEUM OF ART

Walnut case, 91"

2006-90-1

Bequest in 2006 by Robert G. Erskine, Esq. Given to him in August, 1961, by Ephraim Tomlinson of Avalon, New Jersey with this note: To my good friend Robert Erskine, "To you Robert—It is with pleasure that I give you this clock, and assuming you would like to know why or how it came to me I am making the following brief statement. My first knowledge of the clock is that it was owned by Ebenezer Roberts of Moorestown, New Jersey, who in his lifetime gave it to his sister Rachel Roberts Sharpless "his baby sister"— my Mary's mother; and on her death in 1899 [sic], Mary's father Caspar T. Sharpless or her brothers Jesse and Allen R. Sharpless (Mary's twin) turned the clock over to Mary—who protected it until her death on March 9th 1954 ^when^ as a part of her estate, by will probated at Mt Holly it came to me—I feel sure Mary would want you to have this clock, as she always approved the careful consideration I gave to matters of this kind, before setting her faith in me overwhelms me--; and yet her frequent comment, "That about like thee to do this" is a cherished memory, + something to try to live up to— With appropriate regards Your Friend, Ephraim Tomlinson."

Thanks to PMA curator Alexandra Kirtley for providing this transcript and following genealogy: "Ebenezer Roberts (1833–1918) and Mary Lippincott Roberts (1832–1927); to his sister Rachel Mary Roberts Sharpless (1845–1900) and Caspar Thomas Sharpless (1842–1922); to their daughter Mary Thomas Sharpless Tomlinson (1875–1954); to her husband Ephraim Tomlinson (1875–1962); by gift to Robert Galbraith Erskine, Junior (1918–1996)."

## 55
## LOCATION UNKNOWN

Walnut case, 94"

Offered but unsold at September 21, 2002, auction at Pook & Pook, Downington, Pennsylvania.

REFERENCES: DAPC. Pook & Pook, September 21, 2002, Lot 32.

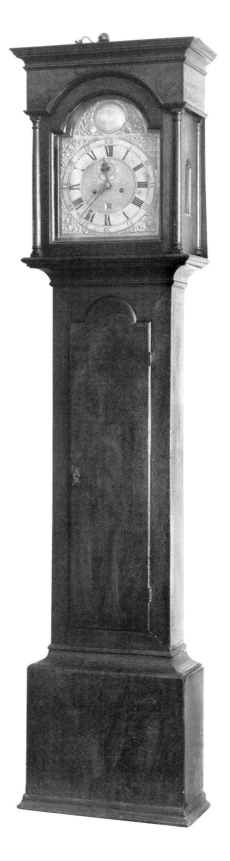

## 56
## LOCATION UNKNOWN

Walnut Case, 88.5" x 20.25" x 10"

Formerly owned by Wilson Antiques, Hickory, Pennsylvania.

REFERENCES: DAPC.

## 57
## LOCATION UNKNOWN

Mahogany Case, 88.5" x 19.75" x 10"

Purchased via private sale in December, 1971, by John F. Hotchkiss (1906–2004) of Rochester, NY, and donated shortly thereafter to the U.S. State Department for the Thomas Jefferson Room, one of the diplomatic reception areas. A later photo showed it standing in the Washington Ladies Lounge. Sold by Christie's June 2, 1990, reportedly to Philip Bradley. Christie's catalogue stated "the works probably not original to the case." Per Hotchkiss obituary: "He was a renowned lecturer, certified appraiser and author of several books on antiques and collectibles. Mr. Hotchkiss was a lifelong member of the Antiques Dealer's Association, the Appraisers Association of America and the Golf Collector's Society."

REFERENCES: DAPC. Obituary John F. Hotchkiss, *The New York Times*, February 15, 2004. "Fine American Furniture, Silver, Folk Art and Decorative Arts," Christie's New York, June 2, 1990.

*Photo courtesy of Christie's*

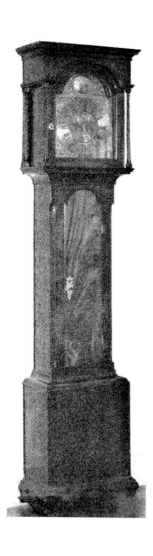

## 58
## LOCATION UNKNOWN

Mahogany case, 90" x 21" x 12"

Formerly owned by Henry Peter Borie of Philadelphia. The Borie family had other Duffield clocks because a Duffield descendant, Patty Duffield Neill (1846–1940), married Beauveau Borie (1846–1929) in 1865. Per Ian Quimby: "Dead-beat escapement. The clock descended through the Borie family from Edward Duffield, their ancestor. It may have been one of those mentioned in Duffield's inventory."

REFERENCES: DAPC. Eckhardt p.285, Fig.3. Quimby 6, October 19, 1962, letter from him to Henry Peter Borie. November 29, 1962 reply from Mr. Borie.

*Patty Duffield Neill Borie*

## 59
## LOCATION UNKNOWN

Mahogany case

James W. Gibbs wrote in a caption that this "probably was made on special order because the case, constructed of Santo Domingan mahogany, has cloth-covered open work to allow a louder sound to emit from the bell."

REFERENCES: Gibbs pp.28–29, fig.22.

## 60
## PRIVATE COLLECTION

Delaware

Mahogany dwarf bombé case

Papers inside the waist door list each step in ownership from Edward Duffield to his son Benjamin and later. Quimby wrote that it came to Mrs. Albert Nalle from Beaveau Borie, Jr., descended in the Borie family as did others in this catalogue, and may be in Edward's estate inventory. A 2020 email to the author from Peter D. Nalle provided family lore about the chain of ownership dating back to the marriage of Beauveau Borie (1846–1930) and Patty Duffield Neill (1846–1940), a Duffield descendant, when the clock entered their family. According to Brooks Palmer, "B. Borie, Whitemarsh, Pa. owns two, one 5' high." An October 24, 1962, letter to Quimby from Henry Peter Borie stated "The third, a "Granny" clock, is owned by my cousin Mrs. Albert Nalle (Patty Borie) of 8828 Germantown Av." A November 7, 1962, letter to Quimby noted "the small tall case clock belonging to my cousin Mrs. Albert Nalle." In 2019 Nalle descendants sold the clock to a dealer who then sold it to the present owner. Not shown is an added scroll top that Quimby dates from 1790-1820. The bombé case form is rare for an American or English clock. The single-train non-striking movement has a deadbeat escapement.

REFERENCES: DAPC. Quimby 5. Eckhardt p.285. Palmer p.184.

*Photos by John Wynn*

## 61
## WINTERTHUR MUSEUM, GARDEN AND LIBRARY

Winterthur, Delaware

English Bracket Clock

Walnut and mahogany veneer, 18" x 10.375" x 6.75"

1958.1929

Bequest of Henry Francis du Pont. Stamped on the movement's front plate is "BEST/No./948" attributed by Winterthur to London clockmaker Thomas Best working in the mid-18th century. Edward was not known to make bracket clocks or to have the ability to fabricate the specialized parts such as mainsprings, crown wheels, and fusees. This confirms that Duffield imported and signed English clocks. Milley referred to a similar table clock signed by another Philadelphian: "Thomas Parker could import clock parts from London and sell the assembled timepiece at a profit after he had put his own name on the dial."

REFERENCES: DAPC. Quimby 14. LaFond XXIX. Eckhardt p.286, Fig.7. Milley p.147. Hummel 1976 p.131.

*Bequest of Henry Francis du Pont, Courtesy of Winterthur Museum, Garden & Library.*

# Edward Duffield Movements, Surveying Compasses, and Sundial

*Photo credits as indicated. Where not shown, courtesy of named auction houses or, if "DAPC", courtesy of Winterthur Museum, Garden & Library: Decorative Arts Photographic Collection.*

## 62
## PRIVATE COLLECTION

Pennsylvania

Movement, hands, and dial only.

*Author photos*

## 63
## PRIVATE COLLECTION

Pennsylvania

Movement (single-train) and dial only.

*Author photos*

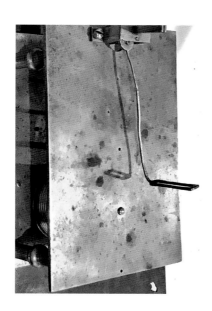

# 64
## THE HISTORICAL SOCIETY OF FRANKFORD

Philadelphia, Pennsylvania

Surveying compass

M 1932.93.2

REFERENCES: "Local Elite", *Northeast Times,* Philadelphia, October 19, 2016. Per Museum Chair Susan Couvreur, the instrument was conserved in 2016 by the late Jeffrey Lock.

# 65
## AMERICAN PHILOSOPHICAL SOCIETY

Philadelphia, Pennsylvania

2019.2

Surveying compass circa 1765 with unusual quartz pivot bearing at needle center.

REFERENCES: *American Philosophical Society News,* Autumn 2019; "Mapping a Nation," APS Museum exhibition April 12–December 29, 2019.

*Photo courtesy of APS*

# 66
## AMERICAN PHILOSOPHICAL SOCIETY

Philadelphia, Pennsylvania

2019.3

Surveying compass circa 1760. Reproduction sight vanes by the late Jeffrey Lock.

*Photo courtesy of APS*

*Jeffrey Lock*

# 67
## MONMOUTH COUNTY HISTORICAL SOCIETY

Freehold, New Jersey

3194 (47)

Undated surveying compass, with wood case, 1947 gift of William Meirs.

REFERENCES: *Asbury Park Press,* October 22, 1978, p.164. *The Pulse of the People: New Jersey 1763–1789*, 1976, p.134.

## 68
## LOCATION UNKNOWN

Surveying compass reportedly restored by the late Jeffrey Lock of Talmadge, Ohio, and sold on eBay with a tripod on December 10, 2001. Dial diameter is 6.785"; overall dimensions are 13.25" by 8.75".

## 69
## PRIVATE COLLECTION

Maryland

Surveying compass in fitted wood box with dial diameter of 6.75". Overall dimensions are 13.625" by 8.25". Purchased at an antique show in York, Pennsylvania, in the 1980's. Restored by the late Jeffrey Lock of Talmadge, Ohio, who reshaped non-original sights to match other known originals.

REFERENCE: Whisker p.236.

*Photo by Gavin Ashworth*

## 70
## PRIVATE COLLECTION

Maryland

Smaller-than-usual surveying compass in fitted wood box made later. Dial diameter is 5". Overall dimensions are 13.25" by 6". Reportedly sold at auction by Freeman's in Philadelphia and then by the late Philip Bradley. Restored by the late Jeffrey Lock of Talmadge, Ohio.

*Photo by Gavin Ashworth*

*Jeffrey Lock*

*Photo by Laszlo Bodo*

## 71
## LOCATION UNKNOWN

Formerly in the collection of the late William K. du Pont. This engraved garden sundial is dated 1757 and is the only known example by Edward. Mr. du Pont reportedly purchased it from a New Jersey woman around 1980. Marked for latitude 40 degrees, it likely was made for Philadelphia which sits at that north-south global position. With highly detailed fine engraving, it includes an additional scale of minutes to add or subtract when converting sundial time to clock and watch time. The time indicated on sundials varies seasonally. This extra feature would be useful for accurately setting timepieces from sundials, and Edward may have made this specifically for that purpose for himself or a customer.

REFERENCES: *The Magazine Antiques,* August, 1992, p.200, article by Donald Fennimore. *Worldly Goods,* p.211.

# INDEX

Academy of Philadelphia  93, 121

Adams, John  59, 101, 103, 126

Affleck, Thomas  58, 59, 103, 175

Alexander Ramsey House  171, 206

All Saints' Church, Torresdale  15, 17, 68, 170

American Philosophical Society  x, xi, 1, 16, 24, 28, 35, 36, 47, 58, 63, 67, 70, 73, 99, 117, 121, 124-126, 164, 170, 171, 174, 176, 181, 182, 199, 236, 237

Ash, Lawrence (clockmaker)  45, 52

Assaying, gold and silver  70

Atlantic County Historical Society  171, 193

Bache, Richard (Franklin son-in-law)  56, 120, 121, 124

Bagnall, Benjamin Jr. (clockmaker)  51, 162

Baltimore Museum of Art  1, 170, 194

Barn at Benfield  2, 109-111

Bartram, John  110, 117

Biddle, Owen (clockmaker)  14, 53, 54, 57, 70, 102, 175, 176

Birch, William (artist)  1, 72, 77, 89, 95, 132

Birnie, Lawrence (clockmaker)  55, 163

Bond, Thomas (scientist)  24, 70, 130

Bordley, J.B. (publisher)  76

Boston Tea Party  99

Boudinot, Elias (silversmith)  59

Braddock Expedition  28, 29, 161

Brahl, Lewis (tenant, goldsmith)  108

Bray Associates and School  xi, 67-72, 76, 118

Bridges, Henry (clockmaker)  62

Brokaw, Isaac (clockmaker)  100

Brown, Gawen (clockmaker)  52

Bruff, Joseph (goldsmith)  53

Carrell, John & Daniel (clockmakers)  60

Census, Federal  105, 110, 112

Chalkley, Thomas (merchant)  50

Chandlee, Benjamin (clockmaker)  50, 176

Clark, Ephraim (clockmaker)  56

Clarke, Richard (clockmaker)  53

Clifton, Henry (cabinetmaker)  59, 175

Clock Sets  55, 60

College of Philadelphia  12, 13, 14, 64, 71

Colonial Williamsburg  1, 55, 170, 198

Columbia University  188

Comly, Isaac (neighbor)  75, 112

Comly, Joshua (neighbor)  31, 112, 138

Compass, surveying  1, 28, 33, 42, 45, 135, 138, 174–176, 236–238

Condy, Benjamin (instrument maker)  103, 176

Contributionship, Philadelphia  67, 69, 106, 107, 109

Coombe, Reverend Thomas  71, 72, 100, 118

Cooper, Jacob (tenant)  108

Cottey, Abell (clockmaker)  50

Court House, Old Philadelphia  77, 91

Crow, George (clockmaker)  60

Cunningham, George (wigmaker)  11, 106

Dawkins, Henry (artist)  98

Decatur, Captain Stephen  110, 128

Declaration of Independence  x, 87, 100, 123, 216

Derham, William (author)  58

Diderot & d'Alembert  62

Dixon, Jeremiah (surveyor)  28, 96

Dowdney, Burroughs (clockmaker)  53, 54, 59, 190

Drinker, Elizabeth (diarist)  67, 100, 103

Duché, Esther (aunt)  11, 74

Duché, Mayor Jacob  11, 74

Duché, Jacob Jr. (scholar)  11, 100, 107

Duffield, Benjamin (grandfather)  7–10, 106, 117

Duffield, Catherine Parry (wife)  7, 10–12, 16, 17, 31, 92, 96

Duffield, Dr. Benjamin (son)  2, 3, 11, 13–17, 76, 82–84, 93, 101–103, 128–130, 179

Duffield, Edward Jr. (son)  2, 12, 15, 17, 18, 71, 73, 110, 112, 113, 115

Duffield, Hannah Leach (step-mother)  10, 11, 106

Duffield, Jacob (cousin)  112, 117, 218

Duffield, Joseph (father)  2, 8, 9, 11, 106

Duffield, Rebecca Potts (daughter-in-law)  14, 16, 179

Duffield, Reverend George (cousin)  11, 13, 30, 97

Duffield, Dr. Samuel (cousin)  16, 70, 73

Dupuy, John (goldsmith, clockmaker)  54

Eckhardt, George H. (scholar)  4, 24, 42, 182, 193, 197–199, 220, 223, 227, 231–233

Ellicott, John (clockmaker)  45, 116, 120

Enslaved People  10, 14, 72, 88, 99, 100, 103, 107, 108, 109, 112, 118, 125

Equation of Time  58, 116, 175

Evans, Lewis (mapmaker)  70

Ewing, John (scientist)  24, 181

Federal Effective Supply Tax  108, 109

Feke, Robert (artist)   10

Ferguson, James (scientist)   35, 45, 75, 116, 121, 122

Flower, Henry (clockmaker)   52, 107

Fox, Joseph (builder)   27, 62, 69, 70, 72, 107

Franklin, Benjamin   Chapter 9, also x, 2, 5, 9–11, 26, 42, 49, 67, 68, 76, 87, 101, 107, 115, 176, 182, 199

Franklin three-wheel clock   45, 46, 116, 119, 121

Franklin, Deborah (wife of Benjamin)   x, 12, 52, 116, 117, 120, 121, 123, 182

Franklin, Sarah (Sally) later Bache (daughter of Benjamin)   15, 100, 117, 120, 121, 123, 124

Franklin, William (son of Benjamin)   2, 70, 116, 118

Franklin, William Temple (grandson of Benjamin)   15, 115, 123, 124, 126

Free Library of Philadelphia   ix, 171, 186

Freemasons   29, 97

Fugio currency (Franklin design)   122, 123

Geddes, Charles (watchmaker)   57

George III, King   2, 98, 120

George Inn   21, 93

Godshalk, Jacob (clockmaker)   62, 108, 175

Gostelowe, Jonathan (cabinetmaker)   13, 31, 101, 103, 117, 175

Gostelowe, Mary Duffield (cousin)   13

Graham, William (watchmaker)   61

Graisbury, Joseph (tailor)   79, 81, 82–85

Grand Federal Procession   101

Hall, David (printer)   23, 116

Hamilton, Dr. Alexander (diarist)   27, 90

Hamilton, Mayor James   27, 96

Harding, James (clockmaker)   38–40, 45

Harrison, John (clockmaker)   63, 116, 118, 120, 121

Head, John (cabinetmaker)   51, 171

Hepburn, Sarah Duffield (daughter)   11, 15, 17, 112, 128–130

Hepburn, Stacy (son-in-law)   11, 15

Hesselius, Gustavus (artist)   3

Hiltzheimner, Jacob (diarist)   16, 67, 72, 74–76, 101, 103, 164

Historic Waynesborough   208

Historical Society of Frankford   236

Holme, Thomas (surveyor)   75, 105, 106

Hopkinson, Francis   11, 12, 64, 67, 70–72, 75, 97, 101, 118, 125–127

Hospital, Pennsylvania   13, 51, 61, 68, 93, 95, 96

Humphreys, James (publisher)   74

Huston, William (clockmaker)   51, 175

Hyatt, John (brass founder)   51

Indentured Servants   13, 54, 60, 61, 88, 90

Independence National Historical Park   106, 170, 211

Ingraham, Edward Duffield (grandson)   3, 16, 19, 69

Ingraham, Elizabeth Duffield (daughter)   12, 16, 17, 19, 129, 130

Ingraham, Francis (son-in-law)   16, 130

Jack, roasting or spit or smoke   10, 32, 57, 65, 134

Jail (Goal), Philadelphia   2, 72, 73, 93, 101

Jefferson, Thomas   63, 64, 87, 101, 123, 230

Jenk, Torben (historian)   76, 108, 170

Kalm, Peter (diarist)   91

Keating, Luke (tenant)   101, 108, 109

Kinnersley, Ebenezer (brother-in-law)   2, 9–14, 67, 68, 117, 121

Kinnersley, Sarah Duffield (sister)   7, 9, 10, 14, 17

Kirkbride, Joseph   73

Laki Volcano   100

Lamb, Andrew (tutor)   92

Leslie, Robert (watchmaker)   62

Levi, Martha (tenant)   109

Lewis, Edmund (clockmaker)   61

Liberty Bell   96, 217

Library Company of Philadelphia   iv, x, 1, 18, 51, 70, 72, 75, 89, 92, 95, 98, 117, 171, 176, 180

Lind, John (Johannes)   31, 32

Logan, James   50, 97, 107, 109, 110, 171

Lower Dublin Academy   2, 15, 27, 75

McCulloch's Pocket Almanac   112, 113

Manor of Moreland   8–10, 17, 110, 123

Martin, William (clockmaker)   50

Mason, Charles (surveyor)   28, 96

Medal, Indian Peace   28–30

Medal, Kittaning Victory   30

Menkevich, Joseph J. (historian)   75, 170

Mercury, transit of   x, 1, 24, 36, 96, 181

Metropolitan Museum of Art   1, 10, 29, 80, 106, 120, 170, 201

Microcosm: Or The World in Miniature   62, 63

Militias, Philadelphia and Pennsylvania   15, 30, 55, 57, 73, 82, 97–100, 107

Mittelberger, Gottlieb (diarist)   79, 92

Monmouth County Historical Society   237

Moraley, William (clockmaker)   61

More, Dr. Nicholas   8

Morgan, Dr. John   12, 13

Morgan, Thomas (watchmaker)   95

Morris, Robert (financier)   68

Munro, Anne (runaway servant)   73

Musical Clock   31, 32, 47, 62, 138

Native Americans (Indians)   18, 19, 29, 30, 68, 76, 98

Neill, Edward Duffield (great-grandson)   7, 8, 15, 18, 19, 27, 117, 126

Nova Scotia   91, 107

Overduff, Jacob (tenant)   112

Paine, Thomas   74, 100

Parker, Thomas (clockmaker)   233

Parry, David (father of wife)   10

Paxton Boys   98, 107

Peale, Charles Willson (artist)   53, 80, 103

Peale, James (artist)   12

Pearson, Isaac (clockmaker)   60, 61

Penn, William   7, 50, 87, 105, 106

Peters, Reverend Richard   3, 11, 12, 68, 69, 70

Peters, Richard Jr.   67, 74, 76

Peterson, Derrick (neighbor)   31, 138

Philadelphia Museum of Art   80, 170, 228

Philadelphia Society for the Promotion of Agriculture and Domestic Manufactures   15, 74, 76

Pinion(s), clock movement   39, 54, 55, 58, 60, 62, 139, 197

Plaster (Plaister) of Paris (gypsum)   76

Plumstead (Plumsted), William   62, 68, 97, 107

Poor Richard's Almanac   96, 97, 116, 175

Potts, Thomas (iron manufacturer)   13, 14, 62

Pottsgrove Manor   14, 16, 171, 179

Randolph, Benjamin (cabinetmaker)   23, 100, 175, 199

Read, John (Franklin brother-in-law)   29, 161

Redman, Sheriff Joseph   68, 70

Richardson, Francis Jr. (merchant)   51

Richardson, Joseph (silversmith)   52, 58, 107, 161

Riley (Reily), John (clockmaker)   60

Rittenhouse, Benjamin (clockmaker)   viii, 42, 60, 175, 176

Rittenhouse, David   ix, x, 13, 14, 22, 28, 32, 36, 38, 47, 58, 59, 63, 64, 96, 98–100, 108, 125, 126, 164, 176

Roach, Hannah Benner (scholar)   105, 107

Rouse, Emmanuel (clockmaker)   53, 54, 59

Royal Society, London   29, 50, 63, 97, 124

Rush, Dr. Benjamin   13, 99, 102, 123, 164

Sauer (Saur, Sower), Christopher (clockmaker, printer)   51

Savery, William (cabinetmaker)   99, 108, 175

Skidmore, Thomas (clockmaker)   60, 61

Smallpox   9, 91, 117

Smith, George (father-in-law)   10

Smith, James (bell caster)   59

Smith, Mary Humphrey Parry (mother-in-law)   10, 12

Smith, Robert (builder)   72, 93

Smith, Reverend William   97, 125, 126

Somer, Martin (tenant)   112, 113

Sommer, Jacob (neighbor)   31, 138

Sprogell, John (clockmaker)   53

Stamp Act   4, 98

State House (Independence Hall) Clock   x, 27, 28, 61, 62, 65, 89

Stedman, Charles (merchant)   62, 68, 69

Stillas, John (clockmaker)   59, 102

Stockwell, David (antiques dealer)   190, 191, 216

Stow, John (brass founder)   59, 96

Strahan, David (printer)   23

Stretch, Isaac (watchmaker)   54

Stretch, Peter (clockmaker)   22, 42, 49–51, 56, 57, 59–61

Stretch, Thomas (clockmaker)   x, 22, 27, 42, 54, 61, 62, 65, 96, 107, 175

Sturgeon, Reverend William   98, 118

Sundial   1, 30, 88, 93, 115, 116, 118, 123, 174–176, 239

Swift, Elizabeth Duffield (sister)   9, 11

Swift, John (neighbor)   13, 75, 112, 128

Swift, Dr. Samuel (brother-in-law)   11–13, 16, 73

Theodolite   29, 176

Venus, transit of   24, 63, 96, 176

Wagstaffe, Thomas (clockmaker)   52, 54, 57, 61

Walton, Aaron (neighbor)   77

Walton, Silas (neighbor)   111, 129, 130

Walton, William (neighbor)   75

Ward, Anthony (clockmaker)   50, 55

Waring, John (Bray Associates)   71, 118

Washington, George   15, 26, 29, 56, 63, 65, 68, 73, 74, 76, 100, 101, 123, 124, 126, 161, 164, 176

Watches and clocks, lost and stolen   24, 26, 27, 56, 57

West, Benjamin (artist)   127

Whitall, John Siddon (merchant)   80

White, Reverend William   11, 15, 16

Whitehurst, John (clockmaker)   7, 35, 45, 46, 116, 119, 120–124, 126, 127

Whitton, Thomas   9, 117

Willard, Simon (clockmaker)   45

William Paca House   216

Winter, John (bell caster)   59

Winterthur Museum, Garden and Library   iv, viii, 1, 4, 7, 22, 23, 54, 58, 131, 170, 175, 197, 233

Wood, John (clockmaker)   22, 42, 52, 58, 59, 61, 92, 107, 175

Wood, John Jr. (clockmaker)   42, 55, 56, 74, 82, 101, 141

Worcester Art Museum   1, 209

Worknot, Martin (tenant)   108, 109

Wyke, John (tool seller)   58

Wyck House   207

Yellow Fever   10, 16, 101, 103

Young Junto   68